# World Forests for the Future

KV-284-942

# World Forests for the Future:

## Their Use and Conservation

**Edited by Kilaparti Ramakrishna**

**and George M. Woodwell**

Yale University Press New Haven and London

Published with assistance from the foundation established in memory of Calvin Chapin of the Class of 1788, Yale College.

Designed by Sonia L. Scanlon

Set in Trump type by The Composing Room of Michigan, Inc.

Printed in the United States of America by Vail-Ballou Press, Binghamton, New York.

Library of Congress Cataloging-in-Publication Data

World forests for the future : their use and conservation / edited by Kilaparti Ramakrishna and George M. Woodwell.

p.   cm.

Includes bibliographical references and index.

ISBN 0-300-05749-0 ✓

1. Forests and forestry.   2. Forest conservation.   3. Forest ecology.   4. Forests and forestry—Economic aspects. 5. Forest policy.   I. Ramakrishna, Kilaparti, 1955–   .

II. Woodwell, G. M.

SD373.W735   1993

333.75—dc20      92-34492

CIP

A catalogue record for this book is available from the British Library.

The paper in this book meets the guidelines for permanence and durability of the Committee on Production Guidelines for Book Longevity of the Council on Library Resources.

10 9 8 7 6 5 4 3 2 1

# Contents

# Acknowledgments

This volume is the culmination of an international workshop on the conservation and use of world forests held in Woods Hole, Massachusetts. Discussions leading to the workshop spanned several months and tapped the experience and knowledge of many individuals actively involved in this field. The editors are most grateful to these mentors and to all of the participants. We would like to thank the following participants for their able contributions to the discussion in Woods Hole:

'Wale Adeleke, Forests Officer, World Wide Fund for Nature, Gland, Switzerland; Vladislav Alexeyev, Head of Forest Ecosystems Laboratory,

Sukachev Institute of Forest and Timber Siberian Division, USSR Academy of Sciences, Krasnojarsk, USSR; Diogenes Alves, Instituto de Pesquisas Especias, São José dos Campos, Brazil; Michael Apps, Study Leader, Ecosystem Modelling and Climate Change, Northern Forestry Centre, Forestry Canada, Edmonton, Alberta, Canada;

Jayanta Bandyopadhyay, Mountain Environmental Management, International Centre for Integrated Mountain Development, Kathmandu, Nepal; Charles Barber, Forests and Biodiversity Program, World Resources Institute, Washington, D.C., USA; F. Herbert Bormann, Professor, School of Forestry and Environmental Studies, Yale University, New Haven, Connecticut, USA; Barbara Bramble, Director, International Programs, National Wildlife Federation, Washington, D.C., USA; Ian Burton, Director, Human and Natural Sciences Interrogation Atmospheric Environmental Services, Environment Canada, Ottawa, Ontario, Canada;

John Cantlon, Chairman, Board of Directors, Woods Hole Research Center, and former Vice-President for Research and Graduate Studies, Michigan State University, East Lansing, Michigan, USA; Rudolf Dolzer, Professor and Vice-President, Mannheim University, and Member, Commission of Enquiry to Protect Earth's Atmosphere, German Bundestag, Heidelberg, Germany; Dan Dudek, Senior Economist, Environmental Defense Fund, New York, New York, USA; Philip Fearnside, Instituto de Pesquisas de Amazonia, Manaus, Amazonas, Brazil; Alan Hecht, Deputy Assistant Administrator, Office of International Activities, Environmental Protection Agency, Washington, D.C., USA; Richard A. Houghton, Senior Scientist, Woods Hole Research Center, Woods Hole, Massachusetts, USA;

A. S. Isaev, Chairman, Forest Committee of USSR, Moscow, USSR; Nels Johnson, Coordinator, Biodiversity Conservation Program, World Resources Institute, Washington, D.C., USA; Nadejda N. Larionova, Expert, USSR State Forest Committee, Moscow, USSR; Alice LeBlanc, Staff Economist, Environmental Defense Fund, New York, New York, USA; Jagmohan S. Maini, Assistant Deputy Minister, Forest Environment, Forestry Canada, Hull, Quebec, Canada; Mary Lou Montgomery, Member, Board of Directors, Woods Hole Research Center, and Conservationist and Community Leader, Falmouth, Massachusetts, USA; Norman Myers, Consultant in Environment and Development, Headington, Oxford, United Kingdom;

Carlos A. Nobre, Senior Scientist, Center for Weather Prediction and Climatc Studies, Instituto Nacional de Pesquisas Especias, São José dos Campos, Brazil; Darrell A. Posey, Instituto Ethnobiologico da Amazonia, Icoraci, Belém, Pará, Brazil; Robert Repetto, Senior Associate and Director, World Resources Institute, Washington, D.C., usa; Salleh Mohamed Nor, Director General, Forest Research Institute of Malaysia, Kuala Lumpur, Malaysia;

William H. Schlesinger, Professor, Departments of Botany and Geology, Duke University, Durham, North Carolina, usa; Ralph Schmidt, Senior Technical Advisor on Forests, United Nations Development Programme, New York, New York, usa; Anatoly Z. Shwidenko, Director, Centre for ussr Forest Resources Assessment 1990 Project, Food and Agriculture Organization, Rome, Italy; Frances Spivy-Weber, Chair, U.S. Citizen's Network on unced, and Director, International Program, National Audubon Society, Washington, D.C., usa;

Ola Ullsten, Ambassador of Sweden to Italy and Chairman, Tropical Forestry Action Plan Independent Review Team, Embassy of Sweden, Rome, Italy; Ann A. Willis, Treasurer, Board of Directors of the Woods Hole Research Center, and Conservationist and Community Leader, New York, New York, usa; Tomasz Wójcik, Forest Research Institute, Ministry of Environmental Protection, Natural Resources and Forestry, Warsaw, Poland; Brooks Yeager, Vice-President for Government Relations, National Audubon Society, Washington, D.C., usa; and Bernardo Zentilli, Senior Advisor on Forests, United Nations Conference on Environment and Development, Conches, Switzerland.

The editors would also like to thank the following people for their interest, advice, and encouragement:

Bert Bolin, Chairman, Intergovernmental Panel on Climate Change, Switzerland; Nitin Desai, Deputy Secretary General, United Nations Conference on Environment and Development, Switzerland; Roland Fuchs, Vice-Rector, United Nations University, Japan; Madhav Gadgil, Professor, Center for Ecological Sciences, Indian Institute of Science, India; José Goldemberg, Minister of Education and Environment, Brazil; Eville Gorham, Professor, Department of Ecology, Evolution, and Behavior, University of Minnesota, usa; Ted Hanisch, Director, cicero, Norway;

Bo Kjellen, Chief Negotiator, Ministry of the Environment, Sweden; Jim MacNeill, Secretary General, World Commission on Environment and Development, Switzerland; Dhira Phantumvanit, Director, Thailand Development Research Institute, Thailand; Sanga Sabhasri, Minister for Science, Technology, and Energy, Thailand; Eneas Salati, Director, National Institute for Amazonian Research (INPA), Brazil; Emil Salim, Minister of State for Population and Environment, Indonesia; Maurice Strong, Secretary General, United Nations Conference on Environment and Development, Switzerland; and Alvaro Umaña, Former Minister of Energy, Natural Resources, and Mines, Costa Rica.

In addition, we have had encouragement and assistance in this project from many colleagues, friends, and organizations. We are pleased to acknowledge the financial support provided by the Environmental Defense Fund, New York, New York; the Homeland Foundation, Laguna Beach, California; the John Merck Fund, Boston, Massachusetts; the Rockefeller Brothers Fund, New York, New York; and the Woods Hole Research Center, Woods Hole, Massachusetts.

In a special note, we commend Julie Williams and Katharine Woodwell for their excellent work in organizing the workshop. We are especially indebted to Julie for her assistance in compiling and editing the manuscript. Working with patience, diligence, and intelligence, she was a great help in the completion of this volume.

Forests rival oceans in their influence on the biosphere. They perform a variety of functions, stabilizing land, controlling flows of water, modulating local climate, providing the major reservoir of land biodiversity, and supplying fiber and even food for people. In fact they play an essential role in the function of the biosphere as a whole, and evidence of global warming has persuaded scientists that grave peril must attend the destruction of the last of the world's grand forests. Yet their immediate value for timber, often very modest, continues to dominate virtually all considerations of management of forested land by individuals, corporate owners, or even governments nominally representing the public trust.

The reasons are many. They include tax policies, policies involving ownership of land, economic exigency, greed, and corruption. They all result, however, in the continuous global destruction of forests, a major factor in the warming of the earth through the release of carbon dioxide and a further contribution to systematic and progressive impoverishment of land and people. Scientists consider such effects serious: already of global proportions, these problems are growing, and many of the effects are irreversible. There is a possibility that the warming of the earth, caused in part by massive deforestation, may soon reach a point where the additional release of heat-trapping gasses that it induces from accelerated decay of organic matter in forests and soils will exceed the extent of control possible through curtailing use of fossil fuels and improving forest management. The warming will feed on itself by destroying forests more rapidly than they can be regenerated. The effect will be a rapid and progressive reduction in the capacity of the earth to support life, including people.

There is no proof of this potential catastrophe. There will be no proof until far too late to deflect the process. But the possibility is real, and in the estimation of scientists the risks are high. The advantages of early action to avoid such disaster would accrue to everyone, but they are obscure and little recognized. One purpose of this book is to explore the role of forests in human life, to start the process of defining the extent to which forests are a part of a stable and wholesome habitat for people.

The chapters contained in the book were commissioned first as background papers for an international workshop held in Woods Hole in October 1991. Planning for the workshop took into account the halting progress of the continuing discussions on forests in the Preparatory Committee of the United Nations Conference on Environment and Development (UNCED). The critical observation that led to these discussions was that in no case has there been political progress in addressing environmental issues internationally without consensus from the scientific community as to the definition of the problem and an equally clear definition of potential solutions.

If governments were effective in their central mission of protecting public as opposed to private interests, such issues as how best to deal with our deteriorating environment would not arise. But governmental purpose is often confused, even corrupted, under relentless pressure from private economic interests that can afford the vigorous pursuit of their own narrow purposes. The public interests become victims of continuous compromise. In the case of essential resources, such as air, land, water, and climate, the effects of the compromises accumulate and threaten a general, global crisis, now upon us in the form of drastic, sudden changes in the human habitat. These changes all contribute to the warming of the earth, but they also include a host of parallel movements whose speed of development exceeds even the supraexponential growth of the human population, which now is expected to double in thirty to forty years. The demise of forests lies at the core of the crisis, along with a half dozen other related issues that have until recently been overlooked in the growth of the human enterprise: increase in population, overuse of fossil fuels as a source of energy, toxification of the environment, progressive biotic impoverishment, a narrowed and narrowing agricultural base, and the destruction of fisheries. All these feed the general tragedy. A solution to one fosters solutions to all; failure to address all assures that none will be resolved satisfactorily.

What do we "need" from forests? How might they be managed to assure their continuity and ensure they provide for the full range of human uses? The complete answer is clearly beyond the reach of any one book, but suggestions lie in the chapters by George M. Woodwell and Richard A. Houghton, who have outlined the importance of forests both globally and locally and attempted to define quantitative ways to stabilize climate. Houghton points out that modern technology allows us to use biomass (plant materials and animal waste) as a substitute for fossil fuels in certain cases and recommends its use. There is a large amount of energy available from that source, and Houghton's discussion shows us that the combination of biomass with the various forms of solar energy available could provide enough energy to displace fossil fuels rapidly and soon.

The suggestions offered in these two chapters, however, do not allow for human growth. If continuous warming threatens human life, so does continued rapid growth in the numbers of people and their enterprises. No change in the way we manage our affairs will open the earth to infinite expansion of human activities. The problems of the moment are the product of an unsustainable growth in activity. Shifting to other patterns of resource use may alleviate our immediate problems, but continued growth in human numbers and technological development would erase those gains in a short time. No magic of law, economics, ecology, or technology can change that fact. Herman E. Daly's call for a century of restoration, discussed below, is appropriate, and Woodwell and Houghton provide suggestions to start the thinking, but the transition needs to be a large one and does not seem to be underway, however urgent.

The discussion contained in Woodwell's and Houghton's chapters goes far beyond the normal limits of economics to emphasize values of forests for what are sometimes referred to as "public service functions," the role of forests in maintaining the general environment. That role is most acutely seen in the case of forest dwellers, whose interests, lives, and history bind them to the forest itself. Their knowledge offers potential advice for the industrial world, according to Darrell A. Posey, who has worked with and studied indigenous forest dwellers of the Amazon Basin. But theirs are cultures that lie outside conventional economics, whose interests are often ignored by the studies of the ecologists as well. Their command of the essential elements of these rich forests, however, exceeds that of the most erudite contemporary scholars. There is obviously an omission here. How does humanity's grand scheme for forests preserve the interests of their indigenous dwellers?

The answer is that it does not, unless that grand scheme incorporates sufficient restraint in the exploitation of the land to ensure the preservation of large tracts of forest intact, tracts big enough to accommodate these cultures comfortably, with whatever degree of assimilation into the industrialized world that they themselves deem appropriate. History has not shown this happening, however, and there is little reason to believe the world is different now.

But if our appraisals of the warming of the earth and its causes and cures are correct, there is ample reason to look for ways to preserve vast tracts of forest intact, starting immediately. What better keepers of the forest could there be than those who are forest dwellers by birth and choice? If Houghton's appraisals are correct, we need forests simply to stabilize the composition of the atmosphere, and we need enough of them to accommodate forest dwellers, indigenous tribes, rubber tappers and their extractive reserves, and parks. And we have as yet said nothing of the control of water supplies and the other ancillary products of forests—equally important uses.

The economists writing in this volume, Herman E. Daly and Robert Repetto, emphasize points that seem obvious, now that they have been defined for us by these eminent scholars. Forests are capital, part of a nation's endowment. Their destruction constitutes not an addition to the gross national product but a loss, a weakening of the capital position of the nation. Daly goes farther, to point out, as ecologists often have, that biotic impoverishment has progressed to the point where we ought to pause globally in our enterprises, in order to allow the reconstitution of the biotic capital of the planet. The call rings true: stopping the massive deforestation currently taking place is clearly necessary, and the process will give new value to living systems and remind us all that we depend on the other life of the planet for our own. Industrialization gives us improved means of exploiting natural resources but must fail as the resources fail. A sawmill, Daly observes, is useless without trees. Relaxing the pressures on living resources that is intrinsic in reducing the use of fossil fuels will be salutary.

The discussion of international issues contained in the chapters by Jagmohan S. Maini and Ola Ullsten and Kilaparti Ramakrishna makes clear the scale of the challenge to the international community. Maini and Ullsten see a number of opportunities arising from a global discussion of the issues concerning forests. Although they consider that the consensus-building exercise then underway in the UNCED preparatory process encouraging, they point out that work on a world forest agreement should begin soon after the adoption of the guiding principles on forests. Ramakrishna finds that forests have not been managed with consistency on even a national level. Globally, even fewer policies have been formulated: forests have rarely even been the sole subject of international treaties. Ramakrishna, after a careful review of international agreements, points out that with the exception of the International Tropical Timber Agreement (ITTA), forest conservation and utilization have not generated any international legal negotiations. Although admitting that the document that has now resulted from the UNCED preparatory process (Agenda 21 and the non-legally binding, authoritative statement of principles) could offer a temporary substitute for a separate treaty on forests, Ramakrishna highlights the need to begin work now by clarifying what is at stake.

Against the backdrop of international discussions within UNCED, and the proceedings of the Tenth World Forestry Congress (see Appendix 1), participants at the Woods Hole workshop discussed over the course of three days the papers that have been adapted for this book. During this time a consensus emerged as to the most appropriate course of action to be pursued in the near to long term on conservation and use of world forests.

Regarding law and policy aspects of the issue the participants felt that there is an urgent need to adopt an international instrument covering the use and conservation of all types of forests that takes into account all of their functions. The moment for such an agreement appears to be now. Efforts in that direction are hampered in part by the lack of consensus among scientists on the fundamentals of the issues. Also creating difficulties is the inability to offer appropriate appraisals of the full costs of any action (including inaction).

A review of international agreements indicates that, with the exception of the ITTA, forest conservation and use have never been the topic of international legal negotiations. Nonetheless, in spite of the desire expressed in different forums, the community of nations is not prepared to negotiate a specific convention on forests. Among the many reasons is the simple one that the questions posed as the basis for this workshop have not been addressed: there is no broad consensus as to the economic or scientific details necessary for the conclusion of an international convention on forest management. While there is every reason to proceed rapidly and globally in the preservation of the earth's remaining forests and the reestablishment of forests over large areas, participants recognized that the process will take a long time and will require a flow of analyses, data, and discussion not yet available around the world. To speed this transition participants in the workshop recommended establishing an international commission to address the global management of forests.

They also concluded that there is an urgent need for an enhanced consensus among scientists on issues concerning sustainable forest management and conservation. Economic analyses of forests and forest management must also be pursued vigorously. Participants felt, however, that searches for scientific and political agreement can and should be complementary. Further analyses in the science and economics of forests and forest management should not preempt or slow continued political progress.

To these ends the participants recommend establishing an international commission on the conservation and use of world forests, with the immediate objective of speeding the convergence of the views of scientists and politicians around the world and identifying the issues for a possible international instrument on forests. A commission would be seen as part of a process to frame a convention on forest use and conservation. Examples of such opinion-forming studies can be found in *Atmospheric Ozone 1985,* the report of an internationally coordinated scientific exercise; the World Commission on Environment and Development's report *Our Common Future;* and the first assessment report by the Intergovernmental Panel on Climate Change (IPCC). The atmospheric ozone panel's report stimulated the political action that resulted in governments taking steps to protect the ozone layer. Likewise, *Our Common Future* solidified support for the Earth Summit—the United Nations Conference on Environment and Development—that took place in June 1992. Adding to this precedent, the IPCC's first assessment report helped generate the needed scientific and political support for the negotiation of a convention on climate change.

The mandate of the commission should include calling upon the scientific community to define the importance of forests in maintaining the human habitat globally and ensuring that these definitions are quantitative: how much forest is there now? How is it changing? Where is it? What must be done to ensure that forests continue to exist and to perform their functions, not only in providing food, fuel and fiber, but also in maintaining the common interest in a human habitat not subject to progressive impoverishment? In addition the commission should be able to see that the global, regional, and local common property values of all types of forests are defined both scientifically and economically and that these values and costs enter into decisions determining the area, structure, or function of forests.

There is general agreement on the urgency of defining and protecting remaining forests globally. This urgency is attached not only to the role that forests have in determining details of the composition of the atmosphere, the stabilization of landscapes, and water flow and quality but also to their role as the major reservoirs of biotic diversity in all latitudes and the habitat of diverse indigenous populations. The information available on both the economic and ecological implications of changes in the area and management of forests, however, is currently inadequate to support the details of more than a general international agreement on forests. The most constructive step the participants in the conference felt that we can take in the early 1990s is to establish the international commission, charging it with addressing all the issues concerned with forests over a period of two to three years, with the specific mandate of stimulating the fusion of ecological and economic interests.

Events within the UNCED and elsewhere since the Earth Summit have not significantly altered the importance of these recommendations. After protracted negotiations in the preparatory committees and at the Earth Summit itself, the nations of the world adopted what is officially known as the "non-legally binding authoritative statement of principles for a global consensus on the management, conservation and sustainable development of all types of forests" (Appendix 2). What the document lacks in terms of specificity is more than compensated for by the fact that it marks the first time that the world community has negotiated and agreed on a set of principles, even ones explicitly termed "non-legally binding." Although as a first step this shows remarkable progress, it falls way short of what is needed and desirable. The principles and relevant portions of Agenda 21 (UN Doc. A/CONF.151/4 [Part I–IV]), also adopted at UNCED, a document notable for its length and what it hopes to accomplish, do leave a window of opportunity to consider the need for and feasibility of appropriate international arrangements to promote cooperation on forest management. (Agenda 21 is the lengthiest document—more than six hundred pages—ever to be negotiated internationally. Split into four sections and forty chapters, it deals with social and economic divisions, conservation and management of resources and development, strengthening the role of major environmental and development groups, and the means of implementing the above.)

A necessary prerequisite for such cooperation is obtaining appropriate answers to the scientific, technical, and socioeconomic issues raised in this book. And forming an international commission on the conservation and use of world forests that draws upon the strengths of similar initiatives appears to be the most practical way of getting those answers. It would also help to form the coalition of interests necessary for a convention on all types of forests. That convention continues to be one of our most pressing needs.

# Abbreviations

TFAP  Tropical Forestry Action Plan
UNCED  United Nations Conference on Environment and Development
UNDP  United Nations Development Programme
UNEP  United Nations Environment Programme
WMO  World Meteorological Organization

# Forests: What in the World Are They For?

**George M. Woodwell**

Ror most of recorded history forests have been recognized not only as a source of personal succor, even wealth, but also as a communal resource, a source of water, game, land, fuel, and timber for all. The early towns of New England, virtually all carved from forests and dependent on them for essential resources, enacted laws that attempted to control deforestation. Many, such as Williamstown in Massachusetts, established town forests that have survived continual exploitive pressures for nearly three centuries. These efforts followed long-standing attempts in England to preserve forests, often as private hunting reserves, more recently as protection for drainage basins. Some of these preserves have survived through the centuries to add grace and stability to the industrial landscape. In fact, a contemporary controversy surrounds the proposed sale of forested tracts in England, set aside until now as drainage basins to provide safe domestic water supplies. The water companies, incorporated as private companies, are now in search of profits and argue that filtration and other techniques make these reserves unnecessary. The sale of the land offers immediate profit to the companies and has the support of developers and politicians. The controversy amplifies the fact that, although the forests were preserved in the interests of safe water supplies, the public sees far more of value than water alone in forested regions.

The coupling of global climatic warming to rates of deforestation and the recognition that forests are the major reservoir of plants and animals on land has brought another, apparently new, set of considerations into world politics. Emphasis centers on the tropics because rates of deforestation are especially high there and the amount of carbon released is large. In addition the public is only now beginning to recognize how extraordinarily rich in life those forests are. Estimates of the number of species on earth have soared as we have begun to discover the complexity of these remarkable communities. Only a few years ago the most aggressive estimates of the total number of species on earth ranged upward to three million. Access to the canopies of trees, an advance in field studies, showed that insects there are far more diverse than had been assumed and estimates of the number of species in the world jumped by a factor of ten to thirty million. Now, with further study of insects and microbes, some scholars believe that there are as many as one hundred million species globally, with by far the largest numbers in the forests of the tropics. Finally, with the destruction of the earth's last giant trees around the world, with the demise of tropical forests clearly in sight, and with the emergence of a host of related problems such as disruption of water supplies, erosion, loss of soil fertility, the demise of fish, birds, and mammals, and the resurgence of pests and human diseases previously controlled, human activities clearly threaten the biosphere as a human habitat.

The issues remain largely inchoate. Emphasis has fallen on "biodiversity," or numbers of species per unit area, in the tropics, where such diversity is high. But forests have a similar role in every latitude: they contain the largest numbers of different kinds of plants and animals of any community on land and might be considered the most highly developed of the terrestrial communities, at least from the stand-point of complexity of structure and the diversity of life and life forms. But they are far more than a simple collection of species, and it is unfortunate that the discussion of biotic resources has been focused on biodiversity rather than on the capacity of the landscape to provide the full range of its biotic potential. When that potential is eroded, as it is constantly, more than mere diversity or numbers of species is lost. Losses extend to the potential of the land to support life and might far better be seen as "impoverishment" than simple loss of diversity.

The extremes of impoverishment following deforestation are informative. Haiti, Madagascar, peninsular India, large segments of semiarid Africa, the Mediterranean basin (including the hills of Attica), Mesopotamia, and now large segments of the Amazon Basin and Southeastern Asia have been systematically impoverished by unbridled and continuous deforestation and overexploitation for pasture and agriculture. Similar but less spectacular and possibly less serious degradation of the landscape has occurred in temperate zones, which are protected in part by younger, more resilient soils and in part by a flora selected through its survival of the disruptions of the climatic changes associated with the glaciations. Deforestation, often touted by governments, land speculators, and others as the route to wealth, is rarely recognized as impoverishment of the landscape and the cause of human poverty. My survey (1990) has defined the process of biotic impoverishment for various terrestrial and aquatic communities around the world in detail.

The connections between the vigor of the landscape and economic welfare are less well defined, especially when the model of a globally industrialized economy is used that contains no clear definition of the absolute dependence of that economic system on biotic resources. Herman Daly's insights into the fallacies in current dreams of continuous economic and industrial growth within a finite biotic realm offer essential guidelines for any such analysis (see chap. 4). There is a crisis now: the expansion of human activities is rapidly reducing the habitability of the earth. One of the most important transitions is the continuous (and probably accelerating) destruction of the last of the world's primary (never previously harvested for timber) forests and with it the impoverishment of the landscape and its inhabitants. Deforestation need not invariably lead to human misery, but the current surge threatens the stability of climate globally and the welfare of millions locally.

Two issues seem important, one political and one technical. The political issue concerns the very purpose of government: protection of the public interests in resources that are shared by public and private sectors. Definition of those interests depends heavily on technical knowledge of what and how large the resources are and how they respond under intensive use, as well as on understanding the threats to public interests that will emerge from excesses or failures in the management of private interests. One of the key problems in protecting the interests of the public is the possibility that democratic systems may not be capable of providing the restraint required either to control growth in demand on resources or to keep the world working as growth in all sectors of the economy continues. The objective of this chapter is to define that problem.

My purpose is to explore in detail the importance of forests and to define local, regional, and global needs for them. I assume that commercial needs for timber, including pulp, are well defined. Our common property rights in the world's forests now emerge as new and important. What are they and how might they be accommodated?

A fundamental issue involved in any consideration of how the world works is whether the model used should be that of the ecologist, in which the primary resource is life itself, or that of the free-market economist, in which the environment is infinitely resilient and the only question is whether we can afford to consider it at all. Not surprisingly, I adopt that of the ecologist and attempt to define within the limits of that model the role of forests in maintaining a human habitat. Economics as commonly defined by economists is largely peripheral to such discussions because the environment is not a part of the market and is given no value.

The most important change in perspective is the recognition that there is a global need for forests that reaches far beyond well-defined commercial interests. That need has two major elements: stabilization of the atmospheric burden of heat-trapping gases, especially carbon dioxide, and the protection of the biotic potential of the land. This potential includes the full complement of species, the inventory and cycles of essential elements, and the normal functioning of the biosphere. Can this newly emergent need be defined quantitatively?

The new global interests must be considered in conjunction with long-recognized local and regional interests in maintaining forested regions for water supplies, as parks and game refuges, for future timber resources, and for protecting slopes from erosion. These interests, which are communal, have always fought with the promise of immediate profits to owners from harvest of the accumulated stock of timber or pulp. The contest continues, never to be resolved until the last vestiges of profit have been wrung from the sale of the capital, and the landscape and the people are left to compete in abject poverty, any profits long dissipated. The issues are not different now. Greed and corruption are important, even major, forces in deforestation in all nations, including the United States. They can be deflected only by the persistent pursuit of truth (basic information on how the world works) and the relentless insistence on constructive purpose and honesty in government.

### The Role of Forests in Global Threats to Climate

The world at present is accumulating carbon as carbon dioxide in the atmosphere at the rate of about $4 \times 10^9$ tons annually. (*Tons* will refer to metric tons throughout this book.) There are two well-recognized sources: combustion of fossil fuels and deforestation. A third source, the warming-enhanced decay of organic matter in forests and soils, especially in the middle and higher latitudes, is now being recognized as potentially large (Woodwell 1983, 1989). Evidence is accumulating that carbon from this source is beginning to have global effects (Houghton 1990; Oeschel et al. 1991; Keeling, Bacastow, et al. 1989; Keeling, Piper, et al. 1989). The size of the new increment is difficult to appraise; it hinges on the warming itself. Success in stabilizing the temperature of the earth will eliminate decay as a major threat; failure to check the warming may speed the release of carbon to the point where it runs beyond the possibility of control. The latter contingency only emphasizes the urgency of effective action in stopping the accumulation of heat-trapping gases in the atmosphere. And to do that requires the immediate removal of about $4 \times 10^9$ tons of carbon from present net annual releases, the approximate annual net accumulation in the atmosphere from 1989 to 1991. This number can be expected to rise, of course, as total global releases rise.

A reduction in releases of carbon by $4 \times 10^9$ tons would have to come from anthropogenic sources that are presumably controllable. Combustion of fossil fuels currently releases about $5.9 \times 10^9$ tons of carbon as carbon dioxide into the atmosphere. Deforestation has not been as accurately measured but probably releases $1-3 \times 10^9$ tons (Houghton 1991). Either or both of these releases might be significantly reduced, to save $4 \times 10^9$ tons in total. A third possibility, reforestation, has the potential for removing carbon from the atmosphere and storing it in plants and soils. The difficulty is that reforestation is directly competitive with agriculture and requires a large amount of land to be effective in storing carbon. An area of $1.5-2.0 \times 10^6$ square kilometers (km²), possibly more, with good soil, is required to support enough forest to store

$1 \times 10^9$ tons of carbon annually. The annual storage will continue as long as gross photosynthesis exceeds total respiration, a period of fifty years or more in many forests. Or it will continue until the forest is again cut, whichever comes first.

A solution to the continued accumulation of heat-trapping gases in the atmosphere requires a reduction in the use of fossil fuels, control of deforestation, and efforts at reforestation. To avoid having to reduce use of fossil fuels, efforts may be made to control the entire problem by better management of forests, but such efforts will be inadequate. The problem is too large to be controlled without reducing current use of fossil fuels in the developed world and avoiding an expansion of use in developing countries. Management of forests is secondary, but vital, as part of the solution.

We assume with the authors of several reviews (WMO/UNEP 1986; Jäger 1988; Houghton et al. 1990; Leggett 1990; NAS 1989) that despite the self-defeating opposition of the current United States administration the nations of the world will move rapidly to stabilize the temperature of the earth by reducing the releases of heat-trapping gases from all sources, including deforestation. If so, they will establish one limit on the minimal global area of forests: no further deforestation can be allowed beyond, say, what had occurred by 1990. Unfortunately, the area of forests remaining in that year is not well known. Michael Jacobson (personal communication, 1991), using data of the Food and Agriculture Organization (FAO) (1988) and the Economic Commission for Europe (1985), estimated that the total area in forest in 1980–1985 was about $3.6 \times 10^9$ hectares (ha). Adding "wooded areas" brought the total globally to $5.3 \times 10^9$ ha or about 40 percent of the land area. Other estimates are generally within this range.

The possibility exists for reestablishing forests on deforested, even impoverished, lands to reduce the atmospheric burden of carbon dioxide. To remove $1 \times 10^9$ tons of carbon annually and store it in forests and soils would require a successional forest with trees in vigorous growth on an area of $1.5–2.0 \times 10^6$ km$^2$). If the land is impoverished, a larger area would be required. The area involved, probably at least $2.0–3.0 \times 10^6$ km$^2$ for $1–2 \times 10^9$ tons of carbon annually, would require a 4–6 percent increase in the forested area of the earth above what was thought to exist during the early 1980s. Forest on such an area might be expected to continue to store carbon at that rate for fifty years, possibly longer, before total respiration would rise to balance total photosynthesis. At that point the forest would have to be retained as forest or the carbon stored would be released again to the atmosphere. A tract of such size ($1 \times 10^6$ km$^2$) is 1000 km $\times$ 1000 km or about 600 mi $\times$ 600 mi. Such tracts are difficult to find and would clearly interfere with agriculture and other uses of land. And as the forest matured, it would attract the same pressures to harvest the trees for profit that have led to the current circumstance.

The possibility of reestablishing and maintaining forests on such a scale has been explored by R. A. Houghton (1990). His analyses, supported by those of Grainger (1988), suggest that in Africa there might be as much as $3 \times 10^6$ km$^2$ of once forested but now degraded land. Latin America and Southeast Asia each have at least $1 \times 10^6$ km$^2$. Houghton estimated that reforestation of these degraded lands and others totaling $865 \times 10^6$ ha might result in the total storage of $150 \times 10^9$ tons of carbon or an average of $1.5 \times 10^9$ tons of carbon annually throughout the next century. Those reforested regions would have to remain in forest, of course, neither burned nor cut for timber, to retain that carbon. These changes in practice would be in addition to a cessation of further deforestation of existing forests.

Much of the land now in "forest" supports stands that are reduced in volume as a result of heavy harvesting, slash and burn agriculture, and other disturbances. These stands can be allowed to recover, thereby replacing in forests some of the carbon now in the atmosphere. Satellite data and modern sampling techniques (Woodwell 1984; Botkin and Simpson 1990) now allow an appraisal of the magnitude of the amount of carbon to be stored, based on better knowledge of the area of forests and the standing stock of trees.

Houghton (1990) has pointed out that biomass can be substituted for fossil fuels in a wide variety of circumstances, and the substitution holds potential for reducing net releases of carbon into the atmosphere. He estimated that the area of forest required would be at least $5 \times 10^6$ km$^2$ or about 14 percent of the forested land of the earth. Such an area was estimated on the basis of FAO/UNEP data (1981) as about one third the area of croplands globally and one quarter the 1980 area of tropical forests. Questions concerning both the efficiency of using energy from this source rather than fossil fuels and the sustainability of yields render such an estimate tenuous. It is nonetheless true that far more energy is available from green plants globally than we now obtain from fossil fuels, and that the renewable use of plants as the primary source of energy will not change the composition of the atmosphere.

A significant reduction in the accumulation of heat-trapping gases in the atmosphere will require both intensive reforestation of large areas and systematic, large reductions in use of fossil fuels. The human enterprise has passed the point where the energy required to support it in its present form can be obtained entirely through green plants. Biomass may provide a partial substitute for fossil fuels, but a variety of other sources will be required.

### Biodiversity and Biotic Impoverishment

If the rate of warming experienced globally during the last decade (0.2°C) is sustained or accelerates throughout the next, we can expect changes in temperature in the middle and higher latitudes of 0.4–0.5°C or more per decade. Such changes are well beyond any experienced in historical time and beyond the limits of global changes that we know about.

A 0.5°C change in mean temperature in the middle and higher latitudes is equivalent to moving latitudinally 50–75 km. The implications are profound because plants and animals have been selected over many generations for the special circumstances of their current habitat. Changing conditions abruptly makes each organism suddenly maladapted . . . and progressively so as the warming continues. Even though a plant species may occur over several degrees of latitude, the individuals from different parts of that range are far from identical: each is part of a subpopulation of the species that has been selected over several, perhaps many, generations for the particular characteristics required to survive in a special segment of the range. These are long-established principles of ecology, no longer in question (Clausen et al. 1940; Ledig 1991). Climatic changes that are more rapid than the generation time of the species may well destroy subspecies throughout the species' range, not merely at the warmer and drier margin. Vulnerability appears in various forms, but commonly in increased susceptibility to insects or disease, and the morbidity is ascribed to these secondary symptoms (NAS 1989). The process is insidious, difficult to measure but cumulative, and can best be described as systematic impoverishment (Woodwell 1970; Woodwell 1990). Carried to an extreme, impoverishment causes the forest to be replaced by shrubland or, later, by grassland or other still more impoverished communities. This pattern of change has been confirmed for forests from widely different habitats around the world (see several papers on forests in Woodwell 1990).

The shift from forest to lesser-statured vegetation involves a release of carbon into the atmosphere as carbon dioxide or methane. The magnitude of the release depends on the area affected, the magnitude of the change in stature, and the extent to which the change affects soils. Most easily measured is any change in the area of forest: deforestation can usually be identified and measured directly on satellite images. Rapid deforestation, now occurring, constitutes globally a loss of carbon into the atmosphere that is commonly thought to be second only to the release from combustion of fossil fuels. Estimates for the current release range from 1 to $3 \times 10^9$ tons of carbon annually (Houghton 1991), mostly from deforestation in the tropics. Increasing pressures on forests globally, especially on the last of the grand trees in the northern hemisphere, are bringing a surge in deforestation there as well. This surge remains unmeasured, despite its importance.

More difficult to measure, or detect, is the incremental impoverishment of forests and other vegetation. Such impoverishment is widely recognized as common, the result of chronic disturbance, including air pollution and the acidification of rain (NAS 1989; Bormann 1990). The area involved in such incremental degradation may be very large, and result in large further releases of carbon. There is no easy approach to making such measurements. Areas that can be used for comparison as "controls" may not be available. In addition, direct effects from chronic disturbance are confounded by secondary infections by disease or insects, and the causes of the changes in the structure of the forest remain as obscure as the fact of the impoverishment. A rapid warming falls into this category: a cause of impoverishment likely to be obscured by secondary changes that become the immediate and obvious causes of morbidity in trees and other plants. Satellite imagery offers no easy basis for measurements, at least in the short term, although efforts are being made to use the higher resolution imagery of Landsat and SPOT to determine the vigor of local populations of trees in the eastern deciduous forest (NAS 1989; Burke, personal communication). These efforts are in an early stage and moving slowly. They offer no immediate application in the context of this discussion.

While measurement is essential, correction of the trend is even more important. Changes in climate at the higher rates considered possible for the next decades will outrun the capacity of plants and animals for adaptation or migration and will cause a surge of biotic impoverishment unimagined in its severity and consequences for the habitability of the earth. The details of those changes remain to be defined. Avoiding them will require a return to rates of change of historic time measured as 1.0°C per millennium or 0.1°C per century.

The transition from hunting and gathering through grazing cultures to agriculture in the industrial age has brought the landscape progressively into private ownership of small tracts throughout much of the world. One of the most serious aspects of this general trend has been the loss of the natural communities that were the habitats of the indigenous flora and fauna. The first to go were the large animals, some probably driven to extinction by hunters who used fire and may have changed the habitat suddenly. Less conspicuous losses included plants. We have lost to the plow and to grazing and other human disturbance the places and conditions that gave rise to the world's great grain crops of Europe, the Americas, and western Asia. We shall never know the seriousness of that loss. At the same time, few argue that the loss of the wolf, the mountain lion, or the eastern woodland buffalo from eastern North America creates a serious contemporary burden, although we have very little basis for judgment of what eastern North America might be like if we had started our settlement of it with the precept that we would assure the preservation of all species. We do not have the luxury of reversing earlier errors that led to extinctions; we can only work from the present.

The dissection of the landscape, however, continues. The process must be recognized as leading to the progressive loss of indigenous biotic potential, the loss of the structure of natural communities, the loss of the populations of plants and animals that have been selected over generations to fit a particular set of environmental circumstances, and the encouragement of small-bodied organisms with short life cycles as opposed to large-bodied organisms with long life cycles. This general process of biotic impoverishment is a worldwide problem. It is compounded by gross and pervasive changes in cycles of nutrients, such as nitrogen and sulfur, and by the destabilization of landscapes as well as species. The problem is global, largely irreversible, and economically expensive to nations. Haiti, denuded of forest and other vegetation and probably the most impoverished landscape in the world, is but one. The question is how to stop the process in a world in which demands for food and other resources in support of an expanding population already exceed supplies; where human welfare, even survival, are often advanced as the reason for allowing further destructive exploitation. The arguments rarely vary, whether the demand is for more land for grazing in the forests of the Amazon Basin or for more Douglas fir for sawmills in the Cascades of the Pacific Northwest: the immediate needs of a few are advanced over the long-term welfare of the region, the nation, or the world. The larger issues of the rapid, progressive biotic impoverishment of the region are rarely argued. Yet these problems are real and becoming acute in the short term, especially in the forested regions of the tropics where hopes advanced by politicians for the resettlement of excess peo-

ple from cities or other agricultural regions have proven misguided within the period of a very few years. Advice of scholars well in advance (Goodland and Irwin 1975) was ignored.

## Global Interests in Forests

The discussion above emphasizes that concepts of "ownership" of forests are changing. The change is occurring around the world and will accelerate over the next few years. The major shift is the recognition that the public has a vital interest globally in forests as an essential element in the human habitat. Current or historical ownership matters little. Recognition of the public interest in forests is not new. What is new is the realization that current trends threaten the continued existence of forests and that their loss implies an immediate, irreversible destabilization and impoverishment of the human circumstance globally. The time when forests were large in proportion to the demands placed on them has passed; forests are now threatened around the world by direct harvest and displacement by agriculture, by an increase in the frequency of fires, by the spread of industrial toxins, and by the rapid warming of the earth. No forest has escaped human influence. Worse yet, the destruction of forests is contributing to the accumulation of heat-trapping gases in the atmosphere, to the acceleration of global warming, and to the further disruption of living systems at the time when human activities and demands are increasing rapidly without evidence of any limit.

Forested lands are held both privately and publicly in the United States and in much of the rest of the world. Ownership has in general been interpreted as conveying the right to determine details of management, including the right to sell the trees for harvest, to clear the land for agriculture or for real-estate development, or to preserve the forest for future needs, such as hunting or personal or public pleasure. Exceptions exist in the Scandinavian countries where public interests in forests have long been recognized and management is closely regulated. Publicly held forests suffer the same types of pressures for exploitation as do privately held ones . . . and governments commonly yield. The common view has been that interests in forests are local, primarily economic, and that the "owner" can make decisions on the use of the land. Although other values may be given lip service, the value that is commonly seen as paramount is timber, whose exploitation vitiates most other values. That pattern is being challenged more effectively and seriously than ever before by the recognition that forests are threatened around the world and their further loss will create unacceptable public, global burdens. The ownership of forests is shifting away from individuals, even away from nations, as we recognize that the common welfare of nations is tied to biotic resources that are threatened by current patterns of management.

We find emerging here the classic issues in management of a commons, best defined by Garrett Hardin (1968). The example Hardin used was the common pasture once available for public use. Each citizen could use the pasture, thereby claiming a piece of communally held property for his own. A citizen who pastured two cattle claimed more than the citizen with one. Although the yield of the pasture might be diminished by the extra use, the reduction was shared by all the users. Restraint in such a circumstance works to the advantage of all; but the individual who adds to his herd gains at the expense of common interests. The result is the inexorable destruction of the resource, the "tragedy of the commons." The solution lies, of course, in some regulatory system invoked to control the use. Governments are formed for that purpose, among others.

The principle applies to all resources held as common property: air, water, land, fish and game, the use of roads, public services, the oceans, parks, or a hundred other such resources we have difficulty recognizing until our access to them is challenged. We seek in governments rules to establish equitability in access while protecting the common interests from excesses. Governments often fail this challenge, partly from ignorance, often from corruption. But the central principles hold, for the individual pasturing cattle on public land in the North American plains and for the corporations that compete to use air and water for corporate purpose, focussing profits on themselves and dumping the costs as wastes into the public realm. They hold also among nations that compete commercially or otherwise to enrich themselves or their clients by exploiting natural resources, whether the resources are fisheries, the last of the whales, the sea floor, the Antarctic Continent, or the forests within their national boundaries.

There is a further consideration: "Sic utere tuo ut alienum non laedas," an ancient precept, well established in English law. "Use what belongs to you in such a way as not to interfere with the interests of others." As people and competing interests accumulate in a finite space, this version of the golden rule slips further into the background. It remains, however, a touchstone of Western law and a basic challenge to all.

Suddenly, as the world shrinks under the continuous expansion of population and the human capacity for exploiting its habitat, the actions of nations within their borders in addressing what have hitherto been seen as strictly domestic issues affect neighboring countries and the welfare of all. Three such issues are before the international community at the moment: the use of chlorofluorocarbons (CFCs) and their potential for destructive changes in the atmosphere that will increase exposure to ultraviolet radiation at the surface of the earth with great detriment to human interests; the need for control of heat-trapping gases, especially carbon dioxide, in the atmosphere to prevent potentially devastating climatic changes; and a newly recognized need for an international agreement governing forests in recognition of the role they play in controlling the composition of the atmosphere, the biotic potential of the land, and various other aspects of the human habitat.

### What Might Be Done?

A casual review of what goes on around the world in management of resources would suggest that there are no rules, no general principles to guide governments or planners, and that we are condemned again and again to fail, until we have allowed ignorance and greed, incompetence and narrow political purpose, to generate a global Haiti, reducing all to impoverishment and squalor. The alternative way to exploit public resources is what Hardin (1968) recognized correctly as mutual coercion mutually agreed upon: governmental regulation to protect the common interest in a viable landscape from the private desire for immediate gain at public expense. Such a transition, however unpopular as political philosophy in a neo-conservative world, offers the only alternative to chaos. It will require judgment as to how to make the world work. Some believe that the world will run itself if we allow it to and will maintain a suitable human habitat in the process. David Ehrenfeld (1991) has written eloquently on this very point. He has issued a plea that we avoid the assumption that we are capable of "managing" a worldly system when we must know we cannot, having failed so miserably and consistently in the management of far less complex issues. The key, he emphasizes, is in arranging to have the world run itself, something the biosphere has been doing for all of time, until we came along to disrupt its normal function to our own clear disadvantage.

We seek some elementary rules, guidelines for keeping the biosphere operating in spite of the large numbers of people who want to live comfortably within it. The answer, if we listen to Ehrenfeld, lies in restraint: avoid becoming so large an influence on the normal functions of the world that repairs become necessary. How large, or small, is that? Clearly, we are too large now, when we are causing the warming of the earth, changes in global cycles of energy and mineral elements essential for life, and a global wave of biotic impoverishment that is closely tied as cause and effect to the other disruptions.

Let us speculate on how we might begin. If we follow our model, we might decide that to keep responses simple and potentially within the realm of political possibility, we might plan to keep large areas of the landscape intact. We recognize that the issue comes down to planning—overt, deliberate planning—to limit human intrusions on commonly held resources. These intrusions are always made in response to immediate self-interest. They can be limited in response to the same self-interest, realized over a longer time. The intrusions are comprehensive: they apply to patterns of the use of air, land, and water. But let us start with land. The decisions we make may seem arbitrary, at least in the first stages, but their results should offer the necessary justification. The primary decision must be to preserve the flora and fauna of the region. We recognize that our objective must be the preservation not simply of species but of the pools of genes that have been developed over generations within the species that adapt each population to its region. But to take this step requires us to understand that those populations are dependent on the continuity of climate and of the basic chemistry, physics, and biology of the site. Suddenly we realize that we must consider air, water, and climate as well as land. Our objective has established difficult demands, with standards set not by human health but by the needs of living systems in general. Such standards are far more rigorous than those set simply to protect people. They follow from any decision that recognizes the need to avoid further losses of biotic potential and to improve the stewardship of land and forests.

The criteria for establishing what might work, in both a technical and a political sense, are few and weak by comparison with the economic arguments always advanced for further exploitation. It will be impossible to avoid arbitrary decisions—or decisions that appear arbitrary.

We have many questions and few clear answers. What, for instance, is the minimal area needed for a population? The minimal area for the eastern woodland buffalo is not coincident with the minimal area for the eastern bluebird, or for the eastern white pine. How do these minimal areas coincide with the requirements for keeping nutrient elements such as nitrogen on the land and not in the groundwater or in lakes and streams? Answers appropriate for one landscape will not do for another. Countering the persistent weakness of political responses to such issues requires simplicity and clarity of objectives, with compelling reasons for each step. If the problem is as serious as it appears—a rapid global erosion of the human habitat—then Daly's requirement of a century devoted to the restoration of the dominance of biotic processes over human excesses becomes appropriate (see chap. 4).

Some of the considerations that might enter such an analysis follow. Answers will come only over time, with much further study, and they will differ from place to place. The issues may be politically provocative, but they are nothing more nor less than the answers required to resolve current challenges in preserving a human habitat. Local and regional analyses start with the clarity of purpose gleaned from a challenge seen globally. I consider only common property values, not commercial values, although the two are not totally independent.

Global needs call for immediate cessation of deforestation and an increase on the order of 10 percent or more in the forested area of the earth.

This statement seems simple enough, but it demands that existing old-growth stands be preserved, or at least managed to preserve their carbon stocks intact, and that at least $5 \times 10^6$ km$^2$ ($500 \times 10^6$ ha) of land not now in forest be reforested. If both steps could be taken immediately, we would be reducing the emissions from deforestation by $1-3 \times 10^9$ tons annually and removing an additional $2-3 \times 10^9$ tons of carbon annually from the atmosphere into plants and soils. The immediate challenge of removing the excess that accumulates in the atmosphere would have been met, at least for the moment. As the difference in concentrations of carbon dioxide between the atmosphere and the surface water of the oceans declined in subsequent years, however, the rate of diffusion of carbon dioxide into the oceans would also decline and still further reductions in releases would be necessary to keep the atmospheric concentration from rising.

The issue of stopping deforestation and increasing forested area has profound economic and political implications. Progress will depend on anticipating arguments with specific definitions and data: what are "existing old-growth forests"? Which forests are more important, which less ? How do we stop the destruction of forests over large areas by industrial toxification? If the preservation of forests is now important, then steps to stop the loss of forest productivity and to reduce industrially caused increases in morbidity of trees and other plants are also appropriate. What standards will be effective? What constitutes deforestation in the context of this analysis? Does reforestation with plantations constitute "reforestation"? At what point and under what conditions will it be acceptable to harvest trees on reforested lands? Is planning a century in advance possible in a world in which the human population might reach $10-20 \times 10^9$?

The technical challenges are numerous, but possible to overcome. We have the technical capacity, if not the funds, to monitor changes globally week by week. Whether any governmental system can live up to this challenge remains to be seen.

If we stop deforestation, we can encourage a systematic effort to maintain local indigenous flora and fauna in situ by avoiding further incursions into natural forests and by committing areas as substantially untouched parks. Some scientists consider ten percent of the land area to be appropriate, but this fraction is arbitrary, and an objective analysis might raise it and indicate that the purpose requires a more or less continuous "matrix" of natural vegetation in the forested regions. Within that matrix various intensities of use of forest and land occur, including regions preserved intact as parks.

What is the pattern of use of land, which includes "wilderness," "wild" or unmanaged forest, managed forest, agricultural land, and sub-urban and urban land, that will lead to the preservation of the essential qualities of environment, including the genes and combinations of genes of indigenous plants and animals of each region? What is the relationship between the size of an area and its capacity for support of indigenous fauna and flora? Can we establish a minimal size for ecosystems, or is the answer to the minimal size indefinite: the larger the size the more stable the community? The only realistic answer is the continuous matrix.

Or is it possible for large human populations to live on an earth severely depleted in plants and animals as long as energy is cheap and industry is vigorous as some would have us believe?

The quality, quantity, and flows of water in streams, lakes, and ground are heavily dependent on the distribution and extent of forests. What area of forested regions would be committed to control these qualities of the general environment if the landscape were being used renewably? It is possible that the commitment might be of the order of 50 percent or more on this criterion alone.

Deforestation almost invariably speeds erosion from uplands and the loss of nutrient elements into water courses. The losses are most serious in older soils that may not have large base-exchange capacities or any capacity for restoring lost nutrients. What areas are appropriately kept in forest to avoid losses of soil and fertility, which would be accompanied by the pollution of water courses? On many landscapes, especially in the tropics, such circumstances may prevail over much of the region, and we have little choice but to preserve the forest substantially intact. Can these regions be defined and managed accordingly?

In the forested zones local climatic effects occur following changes in the distribution of forests. The effects may extend to sufficient changes in microclimate to prevent the reestablishment of a forest once it has been destroyed.

In many parts of the world tropical forests still supply a source of food and fiber for forest dwellers, who use a surprisingly large fraction of the plants and animals for one purpose or another. The exploitation of the environment is complex but usually at a subsistence level; it lies outside the international economy and is commonly ignored by politicians and economists engaged in national or international economic planning. The formal establishment of "extractive reserves" serving the purposes of the Rubber Tappers in the Brazilian state of Acre constitutes one step in the direction of exploring the extent to which the intact forest can be used to produce a cash income on a continuing basis for forest dwellers and provides defense against those who seek to exploit the forest for timber. The extent to which the experiment can be made successful remains to be seen. Various other innovations are underway in the tropics with the objective of improving yields from landscapes kept under natural or near-natural tree cover.

A casual appraisal of current experience might suggest that a stable landscape in much of the forested region of the earth would be one that maintains 75–90 percent in forest and that establishment of such a ratio would resolve major environmental problems both locally and globally. Large areas of the earth probably meet this standard now; others would benefit from establishing it. But objective data defining the current status of land use are few.

The question of whether such a transition would be appropriate can be resolved with far more objectivity than is apparent at the moment. Resolution will require a heavy reliance on modern techniques, including the satellite imagery now in hand but little used for such purposes and a far greater emphasis on the use of the new capacities brought by computers for handling geographical information. The scientific community has been slow to address these issues; money has not been available and the answers have been unpopular. What is clear now is that current trends in the management of forests are threatening all and that a substantial change in direction is necessary. The change will require leadership from the technical and scholarly community not now forthcoming.

It seems essential that the topic be addressed over the course of at least two years by an international commission with the express charge of refining both the questions and the current answers. The process might take as an example the World Commission on Environment and Development and look toward influencing both the scientific and the political communities over years. Any such transition from local to global management of forests, as now seems appropriate, will surely require that scale of effort.

### References

Bolin, B., B. R. Döös, J. Jäger, and R. A. Warrick, eds. 1986. *The Greenhouse Effect, Climatic Change, and Ecosytems.* Chichester, U.K.: John Wiley and Sons.

Bormann, F. H. 1990. "Air Pollution and Temperate Forests: Creeping Degradation." In *The Earth in Transition: Patterns and Processes of Biotic Impoverishment,* ed. G. M. Woodwell, 25–44. New York: Cambridge University Press.

Botkin, D. B., and L. G. Simpson. 1990. "Biomass of the North American Boreal Forest: A Step Toward Accurate Global Measures." *Biogeochemistry* 9:204–209.

Clausen, J., D. D. Keck, and W. M. Heisey. 1940. *Effect of Varied Environments on Western North American Plants.* Experimental Studies on the Nature of Species 1. Washington, D.C.: Carnegie Institution of Washington, Publication No. 520.

ECE/FAO. 1985. *The Forest Resources of the ECE Region.* Geneva: Economic Commission for Europe.

Economic Commission for Europe. See ECE/FAO.

Erhenfield, D. 1991. "The Management of Diversity: A Conservation Paradox." In *Ecology, Economics, Ethics: The Broken Circle,* ed. F. H. Bormann and S. R. Kellert, 26–39. New Haven: Yale University Press.

FAO. 1988. *An Interim Report on the State of Forest Resources in the Developing Countries.* Rome: Food and Agriculture Organization.

FAO/UNEP. 1981. *Tropical Forest Resources Assessment Project.* Rome: Food and Agriculture Organization.

Goodland, R., and H. E. Irwin. 1975. *Amazon Jungle: Green Hell to Red Desert?* New York: Elsevier Scientific.

Grainger, A. 1988. "Estimating Areas of Degraded Tropical Lands Requiring Replenishment of Forest Cover." *International Tree Crops Journal* 5:31–61.

Hardin, G. 1968. "The Tragedy of the Commons." *Science* 162:1243–1248.

Houghton, J. T., G. J. Jenkins, and J. J. Ephraums, eds. 1990. *Climate Change: The IPCC Scientific Assessment.* Cambridge: Cambridge University Press.

Houghton, R. A. 1990. "The Future Role of Tropical Forests in Affecting the Carbon Dioxide Concentration of the Atmosphere." *Ambio* 19:204–209.

Houghton, R. A. 1991. "Tropical Deforestation and Atmospheric Carbon Dioxide." *Climatic Change* 19:99–118.

Jäger, J., ed. 1988. *Developing Policies for Responding to Climatic Change: A Summary of Discussions and Recommendations of Workshop Held in Villach and Bellagio.* World Meteorological Organization/United Nations Environment Programme, Geneva.

Keeling, C. D., R. B. Bacastow, A. F. Carter, S. C. Piper, T. P. Whorf, M. Heimann, W. G. Mook, and H. Roeloffzen. 1989. "A Three Dimensional Model of Atmospheric $CO_2$ Transport Based on Observed Winds: 1. Analysis of Observational Data." *Aspects of Climate Variability in the Pacific and the Western Americas,* 165–236. Geophysical Monograph 55. New York: American Geophysical Union.

Keeling, C. D., S. C. Piper, and M. Heimann. 1989. "A Three Dimensional Model of Atmospheric $CO_2$ Transport Based on Observed Winds: 4. Mean Annual Gradients and Interannual Variations." *Aspects of Climate Variability in the Pacific and the Western Americas,* 305–363. Geophysical Monograph 55. New York: American Geophysical Union.

Ledig, F. T. 1991. "The Role of Genetic Diversity in Maintaining the Global Ecosystem." In *Proceedings of the Tenth World Forestry Congress,* 71–78. Rome: Food and Agriculture Organization.

Leggett, J., ed. 1990. *Global Warming: The Greenpeace Report.* London: Oxford University Press.

Marland, G., and T. A. Boden. 1991. "$CO_2$ Emissions—Global." In *Trends 91: A Compendium of Data on Global Change,* ed. T. A. Boden, R. J. Stepanski, and F. W. Stoss, 386–389. Oak Ridge, Tenn.: Oak Ridge National Laboratory/Carbon Dioxide Information Analysis Center.

Myers, N. 1990. *Deforestation Rates in Tropical Forests and Their Climatic Implications.* London: Friends of the Earth.

NAS. 1989. *Biological Markers of Air-Pollution Stress and Damage in Forests.* Washington, D.C.: National Academy Press.

Oechel, W. C., M. Jenkins, S. J. Hastings, G. Vourlitis, N. Grulke, and G. Reichers. 1991. "Effects of Recent and Predicted Global Change on Arctic Ecosystems." *Bulletin of the Ecological Society of America* 72:209.

WMO/UNEP. 1986. *Report of the International Conference on the Assessment of the Role of Carbon Dioxide and of Other Greenhouse Gases in Climate Variations and Associated Impacts.* Villach, Austria, 9–15 October 1985. WMO Document No. 661. Geneva: World Meterological Organization.

Woodwell, G. M. 1970. "Effects of Pollution on the Structure and Physiology of Ecosystems." *Science* 168:429–433.

————. 1983. "Biotic Effects on the Concentration of Atmospheric Carbon Dioxide: A Review and Projection." in *Changing Climate,* National Academy of Sciences–National Research Council. Washington, D.C.: National Academy Press.

————. 1989. "Global Climatic Change: Warming of the Industrialized Middle Latitudes, 1985–2050, Causes and Consequences." *Climatic Change* 15:31–50.

Woodwell, G. M., ed. 1984. *The Role of Terrestrial Vegetation in the Global Carbon Cycle: Measurement by Remote Sensing.* SCOPE 23. New York: John Wiley and Sons.

————. 1990. *The Earth in Transition: Patterns and Processes of Biotic Impoverishment.* New York: Cambridge University Press.

# 2

# The Role of the World's Forests in Global Warming

**Richard A. Houghton**

D eforestation has a number of environmentally deleterious local effects, such as soil erosion, rainfall reduction, reduction of the capacity of soils to hold water, increased frequency and severity of floods, siltation of dams, and warmer temperatures. In addition, deforestation results in a loss of food, shelter, and other resources for local inhabitants. These effects are particularly troublesome for the people of developing nations, a great number of whom depend directly on the land for their survival.

The effects of deforestation are also global. They include the irreplaceable loss of species; the conversion of potentially productive land to land with diminished capacity to support either crops, forests, or people; changes in the water cycle, heat balance, and climate of the earth; and the emission of chemically active heat-trapping gases such as carbon dioxide, methane, nitrous oxide, and carbon monoxide into the atmosphere. Here I shall consider deforestation and its effect on the emissions of trace gases and, hence, on the earth's atmosphere and climate. I will look at how much carbon has been, is being, and will be released to the atmosphere as forests are transformed to other types of landscapes by humans, how these releases are determined, and how well they are known. I shall also consider how forests might be managed so as to withdraw carbon from the atmosphere—what area of land is physically available to support forests, and how much carbon might be accumulated if forests were reestablished in deforested areas. And finally, I shall detail what a global monitoring of emissions of greenhouse gases from deforestation and reforestation would entail.

This chapter will emphasize tropical forests, largely because the greatest changes in forests are currently taking place in the tropics. Major shifts may be occurring in the forests of the temperate and boreal zones as well, however. In 1980 the net flux of carbon from deforestation and reforestation was calculated to be almost zero (Houghton et al. 1987; Melillo et al. 1988), with releases of carbon from oxidation of wood products approximately balanced by accumulations of carbon in growing forests. These calculations were not based on firm data, however. The extent of change in the area of forests in China and Russia, for example, is poorly known.

In addition to changes in the area covered by forests are shifts in their stature, or the amount of carbon held in their vegetation and soils. We do not know, for example, whether the effects of acid rain and other forms of industrial pollution are increasing or decreasing the carbon content of European and Russian forests. Forests close to sources of pollution are clearly in decline, but more remote areas seem to be experiencing enhanced growth (Kauppi et al. 1992). Whether this growth reflects the fertilizing effect of the nitrogen in acid rain or is the result of regrowth following logging is unclear. Growing trees always accumulate carbon. To estimate the net flux of carbon between forests and atmosphere, however, one must account for changes in all components of the ecosystem—not only the accumulation of carbon in growing trees but also its loss from the mortality and subsequent decay of trees and from the oxidation of logging debris and harvested products (Houghton et al. 1987).

Changes in fire frequency may also have changed the stature of forests and, hence, the exchange of carbon between forests and atmosphere. Fire suppression in the United States and Canada may have led to an accumulation of carbon during most of this century, but in the last ten years the frequency of fires seems to have increased, releasing additional carbon to the atmosphere. Whether similar changes have occurred in the large boreal forests of the former USSR is unclear.

All of these uncertainties underscore the immediate need to synthesize data on forest area, stature, and growth, and to consider the effects of these changes on the terrestrial carbon balance. Knowledge of current trends in the storage of carbon in forests of the northern temperate and boreal zones would provide insight into the response of forests to further changes underway in the global environment. For the remainder of this discussion, however, I shall consider only tropical forests.

### Deforestation—Changes in Area

Two surveys of tropical forests estimated rates of deforestation for the late 1970s. These surveys were carried out by the National Research Council of the United States (Myers 1980) and by the FAO/UNEP (1981). Although both surveys used satellite imagery, most of their estimates were based on reviews and syntheses of ground surveys and estimates of population size. After adjustment for differences in definition, the two estimates of deforestation are remarkably similar in their assessments for the whole tropics (Myers, $7,340 \times 10^3$ ha; FAO/UNEP, $7,253 \times 10^3$ ha), but they differ for individual regions (table 2.1). The FAO/UNEP estimate is 11 percent higher for Latin America (4,119 as opposed to $3,710 \times 10^3$ ha) and 1 percent higher for Africa (1,319 as opposed to $1,310 \times 10^3$ ha). Myers's estimate is 28 percent higher in tropical Asia (2,320 as opposed to $1,815 \times 10^3$ ha) and 1 percent higher for the entire tropics.

This comparison is based on closed forests only; Myers did not consider open forests. Closed forests are large, dense forests that do not allow sufficient penetration of light for grasses to grow on the forest floor. Open forests, sometimes called woodlands or savannas, have grasses between trees or patches of trees. According to the FAO/UNEP study, deforestation of open tropical forests was about $3,820 \times 10^3$ ha. Their estimate for the total rate of tropical deforestation in 1980 was $11,303 \times 10^3$ ha.

As can also be seen by table 2.1, in the ten years since 1980 estimates of the rates of deforestation in the tropics have increased sharply. According to Myers (1989), the annual loss of closed forests has increased

**Table 2.1 Estimated Rates of Tropical Deforestation ($10^3$ ha yr$^{-1}$)**

| | Closed Forests | | | Closed and Open Forests | |
|---|---|---|---|---|---|
| Reference | Myers 1980 | FAO/UNEP 1981 | Myers 1989 | FAO/UNEP 1981 | FAO 1981 |
| Year(s) of deforestation | 1979 | 1976–80 | 1989 | 1976–80 | 1981–90 |
| Tropical America | 3,710 | 4,119 | 7,680 | 5,611 | 8,400 |
| Tropical Africa | 1,310 | 1,319 | 1,580 | 3,676 | 5,100 |
| Tropical Asia | 2,320 | 1,815 | 4,600 | 2,016 | 3,500 |
| Total | 7,340 | 7,253 | 13,860 | 11,303 | 17,000 |

from $7,340 \times 10^3$ ha in 1979 to $13,860 \times 10^3$ ha in 1989 (a 90 percent increase from 1980). The FAO (1991) estimate (including both closed and open forests) also shows an increase (about 50 percent, from $11,303 \times 10^3$ ha to $17,000 \times 10^3$ ha), but the FAO maintains that some of this may be due to underestimates of deforestation in the earlier period. The FAO acknowledges that "deforestation has accelerated in the tropical moist region as a whole" (1990a), although in some Asian countries it has declined (more because they are running out of forests than because they deliberately decided to reduce the rate). It recognizes that deforestation in the late 1970s was underestimated in some of the larger Asian countries, a change that brings its earlier estimate of $1,815 \times 10^3$ ha (FAO/UNEP 1981) more in line with Myers's estimate of $2,320 \times 10^3$ ha for 1979.

According to the estimates of Myers (1989) and FAO (1991), the *increase* in rates of tropical deforestation lies between 50 and 90 percent. The 90 percent estimate (Myers) may be high because a single country, Brazil, accounts for most of the increase found in his estimates, while studies of deforestation by FAO and others have reported a substantially lower rate. The FAO/UNEP study (1981) projected an average annual rate of deforestation for all Brazil of $1.9 \times 10^6$ ha per year for the period 1980 to 1985. In contrast to this projection, an estimate by Setzer and Pereira (1991) for the Brazilian Amazon alone found the rate in 1987 to be about $8 \times 10^6$ ha per year, more than four times higher than the FAO/UNEP projection. Setzer and Pereira found that about 40 percent of this deforestation, or $3.2 \times 10^6$ ha per year, was in closed forests and about

$4.8 \times 10^6$ ha in cerrado, or open forests. Myers's (1989) estimate of $5 \times 10^6$ ha deforestation of closed forests alone is higher than either of these.

Setzer and Pereira considered their estimate, based on the number of fires recorded by satellite, conservative, but other estimates have been even lower. A joint effort by the National Institute for Research in Amazonia (INPA) and the National Institute for Space Research (INPE; Fearnside et al. 1990) shows a figure of $2.6 \times 10^6$ ha for closed forests in 1989. The study estimates that the average rate of deforestation of closed forests in Brazil's legal Amazonia was $2.1 \times 10^6$ ha per year between 1978 and 1989, increasing between 1978 and 1987 but falling substantially between 1987 and 1989. Thus, the rates in closed forests may have fallen from the $3.2 \times 10^6$ ha deforested in 1987 estimated by Setzer and Pereira to values determined recently by INPE: $1.88 \times 10^6$ ha from 1988 to 1989 and 1.38 from 1989 to 1990 (D. S. Alves, INPE, personal communication).

Some of the variation in estimates of Brazilian deforestation results from errors of measurement and extrapolation; some from the fact that different studies included different areas (all Brazil as opposed to legal Amazonia, for example); and some from real yearly differences, such as changes in Brazilian policy or variation in the length of the dry season and, hence, in the number of successful fires used for clearing. These estimates from Brazil are thought to be among the most accurate measurements of deforestation because they are based on satellite data. Although use of satellite data can be problematic, the method offers an objective and repeatable approach to the measurement of deforestation. Scientific uncertainty of how much of the area covered by forests has changed is surprisingly great, given the importance of deforestation to climatic alteration and the demonstrated ability of existing satellite data to measure change in forest area (Woodwell, Hobbie, et al. 1983; Woodwell, Houghton, et al. 1987; Woodwell 1984; Malingreau and Tucker 1988).

Use of satellite data to measure the transformation of forests to non-forests throughout the tropics would increase the accuracy with which we know these transformations and the accuracy with which we can calculate emissions of carbon. A part of any program to measure deforestation, however, should also monitor the use of land following deforestation. For example, is deforestation permanent or temporary? The major discrepancy between the 1980 estimates of Myers (1980, 1984) and the FAO/UNEP (1981) was their findings with regard to changes in shifting cultivation (Houghton, Boone, et al. 1985). Myers believed that shifting cultivation (in which forest land is cleared—often burned—and then cultivated for a few years, after which new land is farmed) was largely being replaced by permanently cleared land and that the area of fallow was decreasing correspondingly. He estimated that about $10 \times 10^6$ ha of fallow forests (young forests with small trees, recovering from previous agricultural use) were cleared for permanent use (Houghton, Boone, et al. 1985). According to Myers, this annual clearing of fallow was almost as large as the annual deforestation of closed and open forests. The FAO/UNEP, on the other hand, projected that the area of fallow would increase between 1980 and 1985, and that shifting cultivation was responsible for 35, 70, and 50 percent, respectively, of the deforestation of closed forests in tropical America, Africa, and Asia. But the FAO/UNEP survey did not consider the fate of lands once they had been cleared for shifting cultivation.

The distinction between temporary and permanent clearing is important because the two processes release different amounts of carbon to the atmosphere. Some of the carbon released with temporary clearing is accumulated on land again with regrowth of fallow forests. Permanently cleared lands, on the other hand, do not accumulate carbon again. They represent a transformation of carbon-rich to carbon-poor lands and, hence, a permanent loss of carbon to the atmosphere. The amount of carbon released from deforestation also varies with the type of forest cleared. Mature forests hold the greatest amount of carbon per hectare, cleared lands hold the least, and fallow forests are intermediate in carbon stocks. Clearing undisturbed forests does not have the same results for the atmosphere as clearing fallow forests, therefore.

### Changes in the Amount of Carbon Held Per Unit Area of Forest

A comparison of estimates of tropical forest biomass shows that those based on direct measurement (Ajtay et al. 1979; Brown and Lugo 1982; Olson et al. 1983) are generally similar to each other but about twice as large as of those derived from wood volumes (Brown and Lugo 1984). The estimates are shown in table 2.2. Each pair of entries shows the high and low estimate of biomass. For the moist closed forests of Latin America, for example, the two estimates are 176 and 82 tons of carbon per hectare. It is unclear which estimate is more accurate (Brown and Lugo 1984; Houghton, Schlesinger, et al. 1985; Fearnside 1986). The high estimate is based on direct and destructive sampling; that is, weighing trees. The total area sampled by this method is small (< 30 ha; Brown et al. 1989). The low estimate is based on volumes of wood (FAO/UNEP 1981). The

**Table 2.2 Carbon Stocks in Vegetation and Soils of Different Types of Ecosystems within the Tropics (Tons of Carbon per Hectare)**

| | Closed Forests | | | | Open Forests or Woodlands | |
| --- | --- | --- | --- | --- | --- | --- |
| | Moist Forests | | Seasonal Forests | | | |
| Vegetation[a] | | | | | | |
| America | 176 | 82 | 158 | 85 | 27 | 27 |
| Africa | 210 | 124 | 160 | 92 | 90 | 15 |
| Asia | 250 | 135 | 150 | 90 | 60 | 40 |
| Soils[b] | | | | | | |
| All tropics | 100 | | 90 | | 50 | |

Source: This table appeared in a slightly altered form in Houghton 1991c. Copyright © 1991 Kluwer Academic Publishers. Reprinted by permission of Kluwer Academic Publishers.
[a]The first value of each pair is based on destructive sampling of biomass (from Atjay et al. 1979; Brown and Lugo 1982; Olson et al. 1983); the second value is calculated from estimates of wood volumes (from Brown and Lugo 1984; Houghton, Boone, et al. 1985).
[b]The values are averaged from estimates in Brown and Lugo 1982, Post et al. 1982, Schlesinger 1984, and Zinke et al. 1986.

areas surveyed are much larger (probably thousands of ha), but estimates of wood volumes have to be converted to total carbon stocks, or the amount of carbon in vegetation (Brown and Lugo 1984; Palm et al. 1986). Recent analysis of the conversion factors has raised the lower estimates of carbon stocks somewhat (Brown et al. 1989), but not enough to equal the high estimates, which are still about 65 percent higher, on average.

Three reasons are thought to contribute to the difference (Houghton 1991b). First, there may still be errors in the factors used to convert volumes of wood into total carbon stocks. Second, the larger area surveyed for wood volumes may provide a more representative estimate of average stocks than the smaller area sampled directly. And third, the stands surveyed for wood volume may have been thinned or degraded. The second and third explanations are linked and seem the most probable. They are consistent with the arguments that the higher estimates of biomass were based on undisturbed sites (Brown and Lugo 1984; Olson et al. 1983). The important question becomes the cause of the disturbance: natural or human. If the overlooked sites were disturbed as a result of human activity, then the high estimates of biomass should provide an estimate for natural forests. And the lower estimates, more representative, will include degraded forests. That degradation is now widespread and increasing in the tropics is clear (FAO 1990a, 1991). Thus, the disparity in estimates of biomass may not be a disparity at all but the result of increasing rates of biomass loss (degradation) from within tropical forests.

## Importance of Forests to the Earth's Climate

Changes in the earth's climate are expected to include a 0.2 to 0.5°C warming per decade and a 3 to 10 cm rise in sea level per decade in the next century (Houghton et al. 1990). The cause of these expected changes is the buildup of heat-trapping gases in the atmosphere. The major gases, besides water vapor, are carbon dioxide ($CO_2$), methane ($CH_4$), nitrous oxide ($N_2O$), CFCs, and ozone ($O_3$). Except for the CFCs, these gases are part of natural cycles among oceans, land, and atmosphere. The increasing concentrations of these gases in the atmosphere, and the enhanced greenhouse effect that results, are due to increased emissions of these gases from human activities, predominantly combustion of fossil fuels and, to a lesser extent, deforestation. Currently, deforestation accounts for about 30 percent of the total enhanced radiative forcing. (Radiative forcing maintains or changes the balance between the energy absorbed by the earth and that emitted by it in the form of longwave infrared radiation.) Carbon dioxide, as a result of both industrial activity and deforestation, has accounted for more than half of this forcing in the past and is expected to account for 55 percent over the next century (Houghton et al. 1990).

The amount of carbon stored in living plants of the earth, about 560 petagrams ($1 \text{ Pg} = 1 \times 10^{15}$ g) is of the same order of magnitude as the amount of carbon stored in the earth's atmosphere (about 740 Pg, mostly as $CO_2$). The amount of organic carbon stored in the soils of the earth is about 1,500 Pg carbon, so terrestrial ecosystems (including both living plants and soils) hold almost three times as much carbon as the atmosphere. Most terrestrial carbon is stored in the world's forests. Forests cover about 30 percent of the land surface, yet they hold about 50 percent of all terrestrial carbon. If only biomass is considered, and soils ignored, forests hold about 75 percent of all terrestrial carbon.

In area, tropical forests account for slightly less than half of the world's forests, yet they hold about as much carbon in their vegetation and soils as all temperate and boreal forests combined. This is largely because of the high biomass of tropical forests. Considering the carbon held in their vegetation alone shows that undisturbed tropical forests hold about 65 percent more carbon per unit area (ha), on average, than forests outside the tropics. Equivalent rates of deforestation in the two regions, therefore, would cause considerably more carbon to be released from the tropics than from regions outside the tropics.

Forests hold more carbon per unit area in trees and soils than the ecosystems that replace them (table 2.3). This carbon is released into the atmosphere as forests are transformed into land that is used for other purposes. The amount of carbon released per unit area depends on the difference between the amount of carbon held in a particular forest and the ecosystem that replaces it; cleared lands may hold twenty to fifty times less carbon per unit area than forests.

Most deforestation releases a large amount of carbon immediately, through burning. Afterward, decay of soil organic matter, slash, and wood products continue to release carbon to the atmosphere, but at lower rates. Once the land stops being used for agriculture, however, regrowth of live vegetation and redevelopment of organic matter in the soil withdraw carbon from the atmosphere and accumulate it again on land. To calculate the net flux of carbon from deforestation and re-forestation, ecologists have documented the changes in the amount of carbon associated with different types of land use and different types of ecosystems in different regions of the world. Annual alterations in the various reservoirs of carbon (live vegetation, soils, slash, and wood products) determine the annual net flux of carbon between the land and atmosphere. Because of the variety of ecosystems and land uses, and because the calculations require accounting for forests in different stages of recovery, bookkeeping models have been developed for the calculations (Detwiler and Hall 1988; Hall and Uhlig 1991; Houghton et al. 1983, 1987; Houghton, Boone, et al. 1985).

The net flux of carbon to the atmosphere from deforestation depends not only on rates of deforestation and on stocks of carbon in forests but on how deforested lands are used. Table 2.3 compares the relative losses of forest carbon as a result of different uses of land. The loss in biomass ranges from 100 percent for permanently cleared land to 0 percent for non-destructive harvest of fruits, nuts, and latex (extractive reserves). Loss of carbon from soil may also occur, especially if the soil is cultivated.

One (non)use of tropical land deserves special attention, from the perspective of deforestation. Most deforestation is for agriculture. Expanding human settlements involves relatively little land, and logged areas—if the logging has not been too destructive and the area is not colonized by farmers—generally return to forests. Despite the fact that

**Table 2.3 Percentage of Initial Carbon Stocks Lost to the Atmosphere from Vegetation and from Soil when Tropical Forests are Converted to Different Kinds of Land Use**

| Land Use | Vegetation | Soil[a] |
|---|---|---|
| Cultivated land | 90–100 | 25 |
| Pasture | 90–100 | 12 |
| Degraded croplands and pastures[b] | 60–90 | 12–25 |
| Shifting cultivation | 60 | 10 |
| Plantations[c] | 30–50 | —[f] |
| Degraded forests[d] | 25–50 | —[f] |
| Logging[e] | 10–50 | —[f] |
| Extractive reserves | 0 | 0 |

Source: Houghton et al. 1987, except where otherwise indicated.
Note: The loss of carbon may occur within one year with burning, or over a hundred years or more with some wood products.
[a]For soils, stocks are to a depth of one meter.
[b]Croplands and pastures abandoned because of reduced fertility may accumulate carbon, but their stocks remain lower than the initial forests.
[c]Plantations may hold as much or more carbon than natural forests, but a managed plantation will hold on average $1/3$ to $1/2$ as much carbon as an undisturbed forest because it is generally regrowing from harvest (Cooper 1982).
[d]From Houghton 1991b.
[e]Based on current estimates of aboveground biomass in undisturbed and logged tropical forests (Brown et al. 1989). When logged forests are colonized by settlers, the losses are equivalent to those associated with one of the agricultural uses of land.
[f]Unknown.

most deforestation accommodates some form of agriculture, however, the agricultural use is apparently not lasting. The annual net increase of agricultural lands remains considerably smaller than the annual net reduction of forest area. For the entire tropics, for example, the expansion of croplands accounted for only 27 percent of total deforestation. Adding the increase in pasture area accounted for an additional 18 percent. Fifty-five percent of the deforestation between 1980 and 1985 was explained by an increase in "other land" (FAO 1990a). Although some of this "other land" takes the form of urban developments, roads, and other settled areas, this probably accounts only for a small percentage of the area deforested. Most of the "other land" seems likely to be abandoned, degraded croplands and pastures, lands that no longer support crop or livestock production, but that do not revert readily to forest, either.

Forests do not convert directly to degraded areas, of course. Conversion is from forest to agriculture, but the agricultural use, in turn, often leads to degradation. My main point is that only about one half of the area of tropical forest lost each year actually increases the area sustainably used for agriculture. The other half is used only temporarily. After a few years it becomes waste, neither agriculturally productive nor forested. If much of the land is being abandoned in this way, making agriculture sustainable may turn out to be twice as effective in halting deforestation as increasing agricultural yields.

The fraction of deforestation used to expand agricultural land rather than to increase the area of unproductive land varies among tropical regions. In Africa the expansion of croplands accounted for only about 12 percent of the net area deforested. Eighty-eight percent of the decrease in forest area was matched by the expansion of "other land." In tropical Asia only 40 percent of the net reduction in forests appeared as an increase in agricultural lands. In Latin America about two-thirds of the reduction in forests could be accounted for by more croplands and pastures. If agriculture could be made sustainable throughout the tropics, rates of deforestation could be reduced by about 50 percent without reducing the expansion of agricultural lands, and large areas of marginal or degraded lands might be reforested.

Deforestation releases carbon to the atmosphere; reforestation withdraws it. The net flux of carbon from the combined effects of both deforestation and reforestation in the tropics was estimated to have been a release ranging between 0.9 and $2.5 \times 10^{15}$ g in 1980 (Houghton, Boone, et al. 1985). Twenty-two of the seventy-six tropical countries contributed 1 percent or more to this total flux; five countries (Brazil, Indonesia, Colombia, Ivory Coast, and Thailand) contributed half of the total net release (Houghton et al. 1987).

No one has calculated the net flux of carbon from temperate and boreal forests since 1980, and an analysis of the current release is overdue. New information indicating significant changes in the areas that have been reforested and in rates of logging has become available.

In the tropics, the net release of carbon after deforestation and reforestation increased significantly since 1980. A 1991 analysis calculated a total possible range of 1.1 to 3.6 Pg carbon per year in 1990, but gave as a more likely range 1.0 to 3.0 Pg carbon (Houghton 1991c). The major uncertainties were rates of deforestation, fate of the land deforested before 1990 (permanent or temporary deforestation), and the biomass of the forests actually cleared.

Part of the carbon emitted into the atmosphere as a result of deforestation is discharged through burning, but the majority is released through decay (Houghton 1991a). On the other hand, the total amount of carbon entering the atmosphere annually from biomass burning exceeds the net release of carbon from deforestation because burning biomass also releases carbon from grasslands, fuelwood, and agricultural waste (Crutzen and Andreae 1990). Thus, the net flux of carbon from deforestation and the gross emissions of carbon from biomass burning are not the same. Using both to estimate carbon emissions inflates the results.

Studying biomass burning nonetheless helps determine change in carbon storage because differences in the frequency of fire affect carbon storage. Reduced frequencies, such as from fire suppression, may increase the amount of carbon stored in biomass. More frequent burning reduces biomass.

Biomass burning is also a major source of atmospheric $CH_4$, CO, and other chemically reactive gases that add either directly or indirectly to the heat balance of the earth. The accumulation of $CH_4$ in the atmosphere contributed about 15 percent of the total radiative forcing in the decade of the 1980s; the contribution from $N_2O$ was about 6 percent. While CO is not itself a heat-trapping gas, it reacts chemically with hydroxyl radicals (OH) in the atmosphere and thereby affects the concentration of $CH_4$.

Emissions of these other gases from biomass burning may be minor compared to emissions of the same gases from subsequent use of the deforested land. Rice paddies, ruminants, and biomass burning are estimated to contribute 20, 15, and 8 percent, respectively, of the total emissions of $CH_4$. Between 50 and 80 percent of the annual release of $CH_4$ comes from terrestrial ecosystems (Houghton et al. 1990). The higher estimate includes releases from natural wetlands and from termites, largely natural sources.

About 65–75 percent of the annual releases of $N_2O$ are thought to come from land (Houghton et al. 1990), with soils alone contributing 50–65 percent. Soils may also provide an important sink for atmospheric $N_2O$, although the magnitude of the soil sink is not known. In fact, we do not understand the global budget for $N_2O$ well enough to account for the observed increase in the atmosphere. Atmospheric concentrations are increasing more rapidly than can be accounted for by the known sources.

A small fraction of the carbon released to the atmosphere during burning of forests takes the form of methane, or $CH_4$. The emissions of $CH_4$ during burning generally lie two orders of magnitude lower than those of $CO_2$, 0.5 to 1.5 percent of the carbon released. The heat-trapping effect of a molecule of $CH_4$, however, is twenty-five times that of a $CO_2$ molecule, so if as much as 4 percent of the carbon were emitted as $CH_4$, the radiative effects of the $CO_2$ and $CH_4$ would be equal in the short term. Because the average residence time of $CH_4$ in the atmosphere is only about ten years, however, while that of $CO_2$ is fifty to two hundred years, the long-term radiative forcing of $CO_2$ is larger than that for $CH_4$.

If the ratio of $CH_4$ to $CO_2$ emitted in fires associated with deforestation is 1 percent, and if 40 percent of the emissions from deforestation in 1990 resulted from burning (Houghton 1991a), then only about 10 Teragrams (Tg, where $1 \text{ Tg} = 1 \times 10^{12}$ g) carbon, as $CH_4$, were emitted to the atmosphere directly from deforestation. But this figure is based on the net release of $CO_2$. Cicerone and Oremland (1988) calculate that gross burning emits 30 to 75 Tg $CH_4$ as carbon ($CH_4$-C) annually, from burning of pastures, grasslands, and fuelwood. In addition, 40 to 70 Tg $CH_4$-C are released from cattle ranching in the tropics and 60 to 170 Tg from rice cultivation (Cicerone and Oremland 1988). Because some of these emissions are from lands that have never been forested, the contribution from deforestation falls somewhere below the total release of 155–340 Tg carbon. In fact, about 35 percent of the global emissions of $CH_4$ may result directly and indirectly from tropical deforestation. The expansion of the wetlands through flooding forests for hydroelectric dams could become a significant new source of $CH_4$ in the future, however.

Nitrous oxide is another biogenic gas that is emitted to the atmosphere following deforestation. Small amounts of $N_2O$ are released during burning, but most of the release occurs in the months following the fire, especially in new pastures. Fire affects the chemical form of nitrogen in soils and, as a result, favors a different kind of microbial activity (nitrification). One of the by-products of nitrification is the production of $N_2O$.

Estimates of the global emissions of $N_2O$ are tentative. Industrial sources are thought to contribute only about 1 Tg $N_2O$-N per year. Earlier estimates of this emission were higher, but the measurements are now thought to have been artificially high. The soils of natural ecosystems are estimated to release 3–9 Tg N annually, as $N_2O$. Fertilized soils may release up to ten times more per unit area, and the soils of new pastures may release even higher amounts. Deforestation for tropical pastures may well be a major contributor to the global increase in $N_2O$ concentrations.

Carbon monoxide is not a greenhouse gas, but it affects the oxidizing capacity of the atmosphere through interaction with hydroxyl radicals, and thus indirectly affects the concentrations of other greenhouse gases such as $CH_4$. Increased concentrations of CO in the atmosphere will deplete the concentrations of OH, leaving less of the radical available to break down the $CH_4$, and thereby increase the concentration and atmospheric lifetime of the $CH_4$. Carbon monoxide emissions generally account for 5–15 percent of $CO_2$ emissions from burning, depending on the intensity of the burn. More CO is released during smoldering fires than during rapidly burning or flaming blazes. The burning associated with deforestation may thus have released 40–170 Tg carbon as CO in 1990. In addition, the repeated burning of pastures and savannas in the tropics is estimated to have released 200 Tg carbon per year as CO (Hao et al. 1990). Together, these emissions from tropical burning equal industrial emissions.

Table 2.4 compares the emissions of greenhouse gases from tropical deforestation and subsequent use of the land with those from other biotic and industrial sources. If we add up the emissions, taking into account the different radiative effects of the gases and their residence times in the atmosphere (Ramanathan et al. 1987), we can see that tropical deforestation accounts for about 25 percent of the heat-trapping emissions globally (see also figure 2.1).

**The Future**

As previously discussed, rates of land-use change in the tropics may have nearly doubled the rate of deforestation over the last ten years. Will the next ten years see another doubling? Will the rate stabilize? Will it decrease? Scientists regularly project the use of energy and the emissions of carbon from combustion of fossil fuels. A logistic curve, however appropriate for the extraction of fossil fuel from increasingly inaccessible deposits, may be inappropriate for projecting rates of deforestation. Forests differ in their accessibility, but the agents of deforestation are ubiquitous and do not require large investments in capital equipment or modern technology. Although such resources are used to clear large areas of tropical forest, tribal shifting cultivators and landless peasants with only primitive means are also responsible for much of the clearing. In contrast to the logistic expression of growth that slows as the residual resource declines in abundance, rates of deforestation may not decrease. Physically inaccessible or legally protected forests may be spared, but even in 1980 forests classified as protected were deforested in most countries within the tropics (FAO/UNEP 1981).

No single factor or group of factors explains variations in rates of

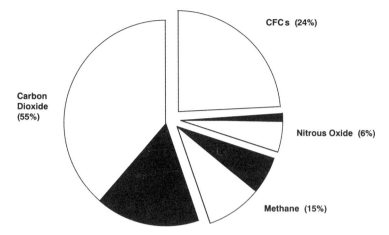

Figure 2.1 Radiative forcing for each of the major human-released greenhouse gases for the decade of the 1980s (Houghton et al. 1990). The shaded portions represent the contributions from tropical deforestation and subsequent changes in use of the land (Houghton 1990b).

deforestation. Deforestation does not appear to be a function of gross national product. The rates are generally higher in developing countries than in developed countries, but there are exceptions to this pattern. Nor is deforestation consistently related to population size. In Brazil, for example, the rate of deforestation for pastures has increased more rapidly than the rate of population growth.

I constructed three projections of deforestation (1990a) by extrapolating rates from the FAO/UNEP data (1981). I used these to calculate the future exchanges of carbon between tropical forests and the atmosphere up to the year 2100. These projections help evaluate a range of possible exchanges of carbon between tropical ecosystems and the atmosphere; they are not predictions.

Between 1980 and 1985, rates of deforestation in closed forests increased by 5.3 percent in Latin America and by 0.6 percent in tropical Asia, respectively. During the same period, rates decreased by 0.2 percent in tropical Africa. My first projection assumes that these rates will continue to increase or decrease linearly to the year 2100. The total rate of deforestation (for all regions, both closed and open forests) will increase from $11.3 \times 10^6$ ha per year in 1985 to $15.8 \times 10^6$ ha per year in 2079, by which time closed forests in Latin America will have been eliminated. Thereafter, the total rate of deforestation will be about $7 \times 10^6$ ha per year. The projection shows a total of $1,518 \times 10^6$ ha of closed and open forests being deforested between 1980 and 2100 (table 2.5).

## Table 2.4 Global Annual Emissions of Greenhouse Gases

| | Annual Emissions | | Percent of Total Emissions | Radiative Forcing Relative to $CO_2$ Molecule | Percent Contribution to Greenhouse Effect in 1980s | |
|---|---|---|---|---|---|---|
| | | | | | Total[a] | From Deforestation |
| $CO_2$ | | | | 1 | 50 | |
| Industrial | 5.6 | Pg C | | | | |
| Biotic[b] | 2.0–2.8 | Pg C | | | | |
| Tropical deforestation | 2.0–2.8 | Pg C | 26–33 | | | 13–16 |
| $CH_4$ | | | | 25 | 20 | |
| Industrial | 50–100 | Tg C | | | | |
| Biotic[b] | 320–785 | Tg C | | | | |
| Tropical deforestation[c] | 155–340 | Tg C | 38–42 | | | 8 |
| $N_2O$[d] | | | | 250 | 5 | |
| Industrial | <1 ?[d] | Tg N | | | | |
| Biotic[b] | 3–9 ?[d] | Tg N | | | | |
| Tropical deforestation | 1–3 ?[d] | Tg N | 25–30 ?[d] | | | 1–2 |
| CFCs | | | | 1,000's | 20 | 0 |
| Industrial | 700 | Gg | | | | |
| Biotic[b] | 0 | Gg | 0 | | | |
| | | | | | 95 | 22–26 |

(1 Pg = $10^{15}$ grams)
(1 Tg = $10^{12}$ grams)
(1 Gg = $10^{9}$ grams)

Source: This table reprinted with permission from Houghton 1990b. Copyright © 1990 American Chemical Society.

[a]From Ramanathan et al. 1987. The greenhouse gases considered in this table are only those released as a direct result of human activities. Tropospheric ozone, formed as a result of other emissions, contributes another 5 percent to the total. The major greenhouse gas, water vapor, is not directly under human control but will increase in response to a global warming (positive feedback).

[b]Biotic emissions include emissions from tropical deforestation as well as natural emissions.

[c]Relatively little of this $CH_4$ is emitted from deforestation. Most of these emissions result from rice cultivation or cattle ranching, land uses that replace forest. Additional releases occur with repeated burning of pastures and grasslands.

[d]Estimates.

**Table 2.5 Projections of Change in the Area of Tropical Forests and Total Net Flux of Carbon to or from Terrestrial Ecosystems between 1980 and 2100**

| | Total Area ($10^6$ ha) | | Terrestrial Carbon (Pg) | |
|---|---|---|---|---|
| Projection | Deforested | Reforested | Released | Accumulated |
| Linear | 1,518 | | 290 (h) | |
| | | | 122 (l) | |
| Exponential | 1,576 | | 291 (h) | |
| | | | 124 (l) | |
| Population | 1,848 | | 334 (h) | |
| | | | 141 (l) | |
| Unmanaged forests | | 865 | | 152 |

Source: This table reprinted with permission of Pergamon Press Ltd. from Houghton 1990a. Copyright © 1990 Royal Swedish Academy of Sciences.
Note: (h) and (l) refer to the use of high and low estimates of biomass. The high estimates (h) include deforestation of $375 \times 10^6$ ha of fallow forests.

In my second projection, the rates of deforestation in 1980 and 1985 were extrapolated exponentially to the year 2100. The results differ little from those of the first projection: total rates increased from $11.3 \times 10^6$ ha per year in 1985 to $20.2 \times 10^6$ ha per year in 2070. After 2070 the annual rate of deforestation decreased abruptly because forests had been eliminated in Latin America. A total of $1,576 \times 10^6$ ha of closed and open forests were deforested between 1980 and 2100.

For my third projection, I assumed deforestation to be a function of the rate of growth of populations, and I based my figures on the expected growth of the population according to the United States Bureau of the Census (1987) and the World Bank (Zachariah and Vu 1988). These sources predict that world population could reach $6.3 \times 10^9$ in the year 2000, $10 \times 10^9$ in 2040, and $13.5 \times 10^9$ by 2100. I increased or decreased the rates of deforestation in each of the three tropical regions for each year depending on the population projection. The rate of deforestation increased from $11.3 \times 10^6$ ha per year in 1985 to about $29.7 \times 10^6$ ha per year in 2045, when the available area of forest in Asia was exhausted. Closed forests in the rest of the tropics were eliminated within the next twenty years; open forests disappeared ten years later. Asian closed forests disappeared first in this projection, while my other two projections show American forests as the first to go. The difference resulted from higher rates of population growth in Asia than in Latin

America. The total area of closed and open forests permanently cleared was $1,848 \times 10^6$ ha.

My rationale for the third projection was that expanding numbers of people will require greater agricultural resources, at least some fraction of which must be provided by an increased area of crops and pastures. Advances in biotechnology and social mechanisms of food distribution may, of course, alter this projection. Grainger (1987), for example, projects that rates of tropical deforestation will decrease by 40 to 80 percent between 1980 and 2020, on the assumption that agricultural productivity will increase. In Latin America the rate of deforestation goes down to zero in one of his projections. Although the technology to support such increases in productivity may be developed, there is no evidence that it will be available to the large and growing number of people outside economic markets, people who can purchase neither food nor technology, but who must use any land available for their own subsistence. Furthermore, as I have already discussed, if rates of deforestation are not to continue to increase, agricultural land will have to be made more sustainable as well as more productive. About half of the deforestation occurring today is to replace worn out or degraded agricultural land. Forests are being converted to agriculture just to keep the area in cultivation constant.

Rates of deforestation vary in my three projections, but the total area deforested does not differ greatly because, once forests are eliminated, no further deforestation is possible. Out of a total area of $1,879 \times 10^6$ ha of closed and open tropical forests left in 1985 (FAO/UNEP 1981), 81 to 98 percent will disappear, according to these extrapolations.

There are two reasons why these estimates are almost certainly too high. First, the projections deal with major tropical regions rather than individual countries. Thus, if the forests of one country were eliminated, rates of deforestation elsewhere in the region need not increase equivalently. Second, we may question whether the forests of a region will ever be eliminated. Both India and China, for example, have reduced their areas of forest to the point where they now have policies to reverse this trend by programs of reforestation. On the other hand, Brazil still feels enormous pressure to provide timber and land from the Atlantic coastal forests, even though only 3 to 5 percent of the original area remains. It is not clear whether the value of forests as forests will ever exceed the value of their timber, land, or other commodities. The possibility that some forests will remain because they are inaccessible is questionable. By 1980, for example, tropical deforestation was occurring in forests designated as legally or physically unproductive for forestry: that is, in forests either legally protected, geographically remote, or physically inaccessible. Such designations have not deterred deforestation in the past, and they are unlikely to stop it in the future if the requirements for land increase in intensity.

Indeed, from another perspective, these projections of deforestation are conservative. My highest projected rate of deforestation for closed forests for 1989 was almost 40 percent lower than the actual rate estimated by Myers (1989).

For each projection, I used two alternative sets of data, to simulate a high and a low estimate of carbon emissions. The high alternative includes the conversion of $375 \times 10^6$ ha of fallow forest to permanently cleared land (Myers 1984). The high alternative also includes the high estimates of biomass for tropical forests discussed in table 2.2, while the low alternative uses the low estimates. (Houghton, Boone, et al. 1985; Houghton et al. 1987).

Estimates in these projections of the release of carbon varied from about 120 Pg carbon for the low estimate for the first two to about 335 Pg carbon for the high estimate in the third projection. These amounts are equivalent to—and up to three times greater than—the amount of carbon estimated to have been released from worldwide deforestation in the 135 years before 1985 (Houghton and Skole 1990). The highest estimate is equivalent to the total amount of carbon that has been emitted to the atmosphere to date from worldwide use of fossil fuels (about 200 Pg carbon).

The annual releases of carbon up to the year 2100, based on the high estimates of carbon stocks, are shown in figure 2.2. The low estimates provide parallel curves, but they are about 60 percent lower per year and overall (see table 2.5). The abrupt reductions in annual emissions after the years 2040, 2070, and 2080 result from reductions in the rates of deforestation, which, in turn, come from the elimination of closed forests from one of the regions. The highest annual release of carbon shown, 5 Pg per year, is nearly the rate at which carbon was emitted from combustion of fossil fuels in 1980.

Rates of deforestation need not continue to increase. The loss of genetic resources, deterioration of local water and land reserves, and modification of global climate resulting from deforestation all argue for stopping or reversing the process. In a fourth projection, I considered the capacity for reforestation to remove carbon from the atmosphere (1990a). This projection was based on a preliminary estimate of the area available for reforestation. My criteria for availability were that the land had supported forests in the past and that it was not presently being used

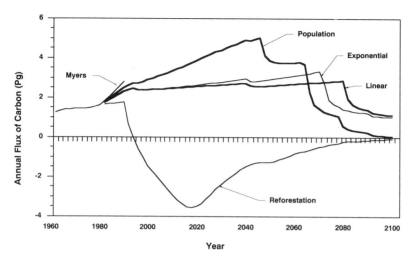

Figure 2.2 Annual fluxes of carbon between terrestrial ecosystems and the atmosphere based on alternative projections of future rates of tropical deforestation and reforestation. Positive values indicate a net release of carbon; negative values indicate a net withdrawal from the atmosphere. The abrupt reductions in emissions in the mid- to late twenty-first century result from elimination of forests in some regions and hence a reduction in deforestation. *Source:* Houghton 1990a.

for croplands or settlements. The first criterion allowed only those regions climatically and edaphically suitable for forests. Reforestation of deserts, for example, was not considered feasible. The second criterion limited the available area to land with little or no current use. With these two criteria, I estimated the area of land that could be reforested for each of the three major tropical regions.

In Latin America the area available, mostly degraded grazing lands, comes to about $100 \times 10^6$ ha. In tropical Asia another $100 \times 10^6$ ha was estimated to be available, largely in degraded grasslands of *Imperata* spp. In Africa I calculated that a total of $300 \times 10^6$ ha were available, more than half of this in the savannas of western Africa. This estimate relies on evidence that large areas of savanna were previously forested and that a combination of fire and human activity has caused and is still causing the savannas to expand. The FAO/UNEP survey (1981) suggests that the area of closed forests in the western regions of Africa may once have extended to a latitude of 10°N: in other words, to an area twice that of the present range of closed forest in western Africa ($190 \times 10^6$ ha). The assumption that this land will support forests again requires that the climate is not, and will not become, more arid than it was when such forests existed.

The total area of previously forested lands not intensively used was thus estimated to be about $500 \times 10^6$ ha when all three tropical regions were combined. This forms about 25 percent of the land currently occupied by tropical forests (FAO/UNEP 1981). Grainger (1988) has calculated there to be a somewhat larger area of degraded lands available for reforestation ($758 \times 10^6$ ha).

A second source of available land is land in the fallow cycle of shifting cultivation. The average stocks of carbon in fallow forests are low compared with undisturbed forest, and the replacement of shifting cultivation with low-input, permanent agriculture has been estimated to allow about 85 percent of this area in fallow to return to full grown forests (Sanchez and Benites 1987). When I applied this percentage to the area of fallow lands available in 1980 (FAO/UNEP 1981), it yielded an additional $365 \times 10^6$ ha for reforestation.

Thus, the total area believed to be suitable for reforestation comes to $865 \times 10^6$ ha. If this is to be reforested over the next century, future increases in agricultural land will have to come from areas neither currently nor previously forested. Clearly, the future of tropical forests depends to a large degree on the demands, yields, and sustainability of food crops (Grainger 1987).

In this fourth projection, I assumed, arbitrarily, that deforestation stopped completely in 1991 and that reforestation of the $865 \times 10^6$ ha occurred over the period 1990 to 2050. According to this projection the lands of previously abandoned shifting cultivation accumulated 54 Pg carbon, and the $500 \times 10^6$ ha of reforested land added 98 Pg. The total withdrawal of carbon from the atmosphere over the period from 1990 to 2100 was therefore 152 Pg. I chose the timing of these activities arbitrarily, but the peak rates of carbon accumulation exceeded 3.0 Pg per year for about a ten-year period around 2020 under the rates projected in figure 2.2. The average rate of accumulation over the century was about 1.5 Pg carbon per year.

Subsequent work using satellite imagery to classify current use of lands in the tropics suggests a much smaller area of land with the potential for reforestation (Houghton et al. In manuscript). When I compared current land cover, determined from the satellite data, with estimates of the distribution of forests and woodlands before human disturbance, I discovered that somewhat less than $250 \times 10^6$ ha of degraded lands are probably available for reforestation. On the other hand, more than $1,000 \times 10^6$ ha of secondary and fallow forests have the potential to accumulate carbon if they are protected from further logging or farming. And about $1,500 \times 10^6$ ha of crop lands and pasture lands could store carbon if some form of agroforestry were introduced. The areas with potential to accumulate carbon are much larger and widespread than the earlier work suggested, although the potential increment per unit area is smaller. Overall, some 100 to 250 Pg carbon might be accumulated in the tropics if management schemes encouraging the growth of woody biomass were implemented.

Reforestation, although useful in the short term to withdraw carbon from the atmosphere, does not represent a permanent solution, however. Once forests have grown to maturity, they no longer accumulate carbon. They continue to hold the carbon they have acquired, but at maturity they are in balance with respect to carbon, neither accumulating nor releasing it. The most effective use of "forests" from the perspective of atmospheric carbon dioxide would be to substitute wood energy for fossil fuel energy. Emissions of carbon would be similar but as long as next year's wood supply was accumulating in growing forests, the emissions of carbon from combustion of wood would be balanced by these accumulations. Fossil fuels, of course, acquire carbon much too slowly to be used sustainably in this way.

I put *forests* in quotation marks above because we may discover that the most effective production of fuel on land is not consistent with the maintenance of forests. Forests provide many more services to humankind than storage of carbon or sustainable provision of fuel. One of the major challenges for scientists is to characterize and quantify the value of forests as forests. Is the best use of forests managing them for carbon or energy reserves? No matter what we decide, stopping deforestation is imperative, both to increase carbon storage and to insure forest protection. One third to one half of the carbon emitted to the atmosphere by human activities comes from deforestation, and about 25 percent of all human-made greenhouse gases result from either deforestation or the land uses that replace forests.

So far this discussion relating forests to greenhouse gases has considered only those changes brought about by direct human intervention. There are other transitions, indirect and inadvertent, that may also be important. For example, arid lands have been expanding in recent decades (Dregne 1985; Mabbutt 1984). Examples include the Sahel, northeastern Africa, northeastern Brazil, and northern Australia. Whether the cause of drought in the Sahel has been population pressure, manifested through overgrazing and the removal of tree cover for fuelwood, or natural variability in climate, the net effect has been to displace populations of people from the Sahel to the northern fringes of forests and woodlands in West and Central Africa, with subsequent loss of forests (Gorse 1985).

The observations suggest that deforestation leads to a drier climate, a finding confirmed by some climate models (Salati 1987; Lean and Warrilow 1989). Unless global changes bring more precipitation to these areas and counteract this trend, more forests may disappear from tropical regions than those destroyed by logging and human clearing. Furthermore, the evidence implies that deforestation will be difficult to reverse. Remove a large area of forest once, and the local environment may no longer support forests at all. Fire seems to be one mechanism for such a feedback. More common in forests occupied by farmers or loggers than in intact forests (Malingreau et al. 1985; Uhl and Buschbacher 1985; Woods 1989), fires may not contribute substantially to forest destruction, but they probably prevent abandoned lands from returning to forest.

There are other reasons to suspect that reductions in forest area could be self-amplifying. One of the effects of a global warming, for example, is likely to be an increased rate of respiration (including decomposition of soil organic matter). Increased emissions of respiratory $CO_2$ and $CH_4$, in turn, would increase the warming. On the other hand, increased concentrations of $CO_2$ in the atmosphere could increase forest productivity and carbon storage. The net effect of the numerous global changes currently underway is difficult to predict. The danger, of course, is that the changes will reduce the capacity of the earth to support life, and, further, that they will initiate processes in the earth's climatic system that are irreversible in human lifetimes. We are unable to say at present whether the current reduction in forest area, worldwide, could initiate or hasten such changes.

### A Blueprint for Monitoring Annual Emissions of Greenhouse Gases from Deforestation

Warming and deforestation are global problems that will require global solutions. One necessity is an international agreement limiting emissions of greenhouse gases. The agreement would have to include an acceptable limit to the rate of global warming. Once this limit was agreed to, scientists could calculate the allowable concentrations of greenhouse gases and, in turn, the allowable emission rates. Whatever allowances, incentives, and penalties are agreed upon for setting these limits, some sort of monitoring will eventually be required to determine annual rates of deforestation and reforestation and to ascertain the quantities of coal, oil, and gas annually consumed by individual countries. In the following pages I outline a program for monitoring emissions of greenhouse gases from forests.

Because carbon dioxide is expected to have the greatest heat-trapping effect (see table 2.4), the first objective must be to figure out changes in the terrestrial stocks of carbon. Emissions of the major greenhouse gases other than carbon dioxide can generally be tied to either biomass burning or land use, which must also be assessed as part of the monitoring program.

In principle, there are at least four approaches for determining these emissions. The most obvious method would be direct measurement of emissions in the atmosphere. This approach is not feasible, however. Measurements would have to be made continuously in time and almost continuously in space. Furthermore, the turbulence of air within the atmosphere makes it difficult to assign measured emissions to specific locations or countries.

A second means would be to inventory carbon stocks on land, including the stocks in trees and soil as well as in paper products, dwellings, landfills, and other storages. Successive annual inventories might ideally provide a measure of the net flux, but the estimate cannot be accurate. Neither would the approach indicate whether carbon emissions were in the form of $CO_2$, $CH_4$, or CO. Nitrogen inventories to determine $N_2O$ flux would be even more problematic because nitrogen can be lost from terrestrial ecosystems in many forms other than $N_2O$.

We could also multiply annual changes in the area of forests by the difference between the carbon stocks of forests and the ecosystems that replace them. This way assumes that decay of organic matter and regrowth of forests are instantaneous. Under a universal monitoring system, a country using this approach might receive all of the demerits for deforestation or all of the credits for reforestation during the initial year of change. In fact, emission of carbon from deforestation generally occurs more rapidly than accumulation of carbon during reforestation; the lags of regrowth are greater than the lags of decay. This simplified approach does not allow for these differences and would not provide accurate estimates of annual flux. In addition, there are several other problems. This approach would require monitoring of regrowing forests to determine that the reforested areas were not deforested again before they reached the specified carbon stocks. The incentive to conserve wood products would disappear because carbon in them would already have been counted as released. Most important of all, the approach requires monitoring biomass within forests. Selective logging and fuelwood harvesting reduce the biomass of forests without deforestation, that is without being "seen," yet the emissions of carbon from these reductions of biomass can be as large as those from outright deforestation (Flint and Richards 1991; Houghton 1991b).

The fourth approach resembles the third in accounting for the fate of all the carbon on a disturbed site, but it includes the time lags introduced to emissions through decay and regrowth. The calculations are therefore more complicated. Instead of multiplying average changes in carbon stocks by changes in area, year-to-year losses and accumulations of carbon in each of the components of the system (living trees, dead plant material, harvested products, soil) are calculated to yield more realistic emission measurements. The data required for this approach are annual assessments of changes in area and carbon stocks. The approach is the same as that described here and used elsewhere to calculate current and historic emissions of carbon (Houghton et al. 1983, 1987; Houghton, Boone, et al. 1985; Houghton, Schlesinger, et al. 1985; Detwiler and Hall 1988). The approach is the most accurate one and the best suited for monitoring emissions.

There are three different, but overlapping, human activities that affect carbon storage in forests. They are de- and reforestation (which change the area of forests); logging, fuelwood harvest, and degradation (which change the carbon stocks within forests); and biomass burning (which releases carbon from other ecosystems as well as forests). Any effective monitoring program must address all three.

To assess the emissions from each of these activities requires in turn that several kinds of data be supplied: identification of how large the affected area was (how large an area was deforested, reforested, degraded, burned); the fate of the affected area (is the effect permanent or temporary?); the land use following deforestation (for example, rice paddy, grazing ground for ruminants); estimate of the initial stocks of carbon in the biomass and soil of the disturbed site; the fate of the initial biomass and soil carbon on the site; and the fate of the initial biomass off the site. Although we need the same kinds of data to monitor each of the three activities, I will note some important differences between the data needed to assess deforestation and those necessary to explore degradation and biomass burning.

The most important component of a monitoring program is measurement of *change in the area* of forests. Satellites can best detect such change, and the best satellite imagery for the purpose probably comes from the Landsat multispectral scanner (Landsat MSS), because of its high resolution and relatively low cost (see table 2.6). A single coverage

**Table 2.6 Resolution, Coverage, and Cost of Imagery from Different Satellites**

| Satellite | Frequency of Overpass | Resolution (size of pixel) | Coverage per Scene (ha) | Cost per Scene ($) | Cost ($ per 10⁶ ha) |
|---|---|---|---|---|---|
| AVHRR | Twice daily | 1, 4, and 15 km² | $90 \times 10^6$ | 130 | 0.7 or less |
| Landsat MSS | 16 days | 0.6 ha | $3.4 \times 10^6$ | 200–1,000 | 60 |
| Landsat TM | 16 days | 0.09 ha | $3.4 \times 10^6$ | 4,400–5,500 | 1,294–1,617 |
| SPOT | 26 days (with 5-day revisit length at ±27°) | 0.04 ha | $0.36 \times 10^6$ | 2,450 | 6,805 |

Source: Lillesand and Kiefer 1979.
Note: Costs are approximate purchase costs (as of January 1992).

of all the world's forests would require about 4,200 Landsat scenes (Woodwell et al. 1983), or about $840,000 for data two years old or older ($4,200,000 for current data).

Such a complete assessment is especially important early in the monitoring program or, ideally, before the program begins. Besides providing a map of forest change, the photographs taken over the first year could be developed into a digital map of land cover. It would serve as a base to measure yearly deforestation. The base map would be a layer of a Geographic Information System (GIS: software for overlaying computerized maps), and subsequent years of complete coverage could be compared against it and against each other to determine change.

We need to make annual assessments of deforestation rates because year-to-year variation in the rates of tropical deforestation is apparently much greater than year-to-year variation in global combustion of fossil fuel. This assessment need not require wall-to-wall coverage with Landsat data, however. It might be based on identification of fires with Advanced Very High Resolution Radiometry (AVHRR) data. The AVHRR imagery of the earth is obtained twice daily. The resolution is coarse (about 1 km² pixels), but the data are relatively inexpensive. The AVHRR data are also available at resolutions of about 4 km² and 15 km². Fire frequency would provide an index of annual deforestation for the dates between those more accurately surveyed with Landsat data. The base map could be used to determine which fires were in forests and which in pastures or other nonforests. For example, INPE monitors fires daily (Setzer and Pereira 1991). Fires monitored from space can be located and visited on the ground in the same day, *while they are burning.* Monitoring of fires also estimates annual emissions of methane and carbon monoxide from biomass burning.

Daily coverage of the earth is only possible with low resolution data (4 km² pixels). There is not enough memory aboard the space craft (or enough receiving stations to unload the data before the memory is full) for full coverage with higher resolution data (1 km² pixels). Daily coverage might be possible with 1 km² data, however, if the areas to be viewed were selected each day for specific regions with intensive burning seasons. Presumably burning seasons vary in different parts of the tropics. The sampling plan would have to be based on the timing and lengths of dry seasons and the frequency of fires. Fortunately, most burning for agricultural land is seasonal and occurs during the dry season, when cloud cover is low and the ground is visible.

Each year's data should also include a 1–10 percent sampling with high resolution, relatively expensive Landsat thematic mapper (TM) or

SPOT (Système Probatoire d'Observation de la Terre) imagery. The sampling would be reserved for areas with active deforestation, logging, or degradation. Sampling or wall-to-wall coverage would also be used to determine the short-term fate of deforested land. How long after clearing and abandonment does a disturbance remain visible from space? As the monitoring program proceeds, knowledge of land-use patterns and dynamics will suggest the optimum frequency of sampling.

The total amount of carbon released to the atmosphere following deforestation and its rate of emission depend on the subsequent use of the land. There are two aspects to this issue. First, will the clearing be temporary, followed by regrowth of the forest, or is it permanent? Second, if the land has been permanently cleared, to what purpose will it be put? Clearing forests for cultivation releases the greatest amount of carbon; careful, selective logging followed by regrowth is among the land uses releasing the least (see table 2.3). Monitoring by satellite can help determine the current use of the land. Detection of changes in forest area will indicate whether the cleared land has started to regrow to secondary forest or remained clear. As already discussed, alternative uses of the land also affect the fate of the carbon originally held in the forest.

Determination of land use is particularly important for estimating emissions of methane and nitrous oxide. As I mentioned above, in addition to biomass burning, the land uses with major emissions of methane are rice paddies and grazing of ruminants. Nitrous oxide emissions are generally greatest in the months following burning or in agricultural fields receiving nitrogen fertilizers. The higher resolution data from satellites can help us identify these land uses.

Two methods may be proposed for obtaining data on biomass and soil carbon. The first calls for using existing data, often approximate, to specify the carbon stocks in forests. These initial estimates, together with monitored changes in forest area, would allow calculation of annual carbon emissions. The second method determines biomass directly, either by ground studies or by remote sensing. Although this plan is more accurate (and more expensive) than the first, it is not yet feasible via satellite imagery, because direct assessment of biomass from space has not been demonstrated.

The first plan, however, also requires surmounting a number of difficulties. Measurements of biomass exist for relatively few forest stands. Surveys of stemwood volumes cover a larger area, but even these surveys leave large areas of the forest unmeasured. Furthermore, as discussed above, high and low estimates of biomass disagree by about 65 percent (see table 2.2). The first plan would thus contain uncertainties of this magnitude. On the other hand, there are modelling approaches being developed that predict the biomass and soil carbon of undisturbed ecosystems on the basis of climatic and soil factors. Such models could reduce the uncertainty of biomass in undisturbed forests but would leave unaddressed the larger problem of the extent of degraded forests.

The second plan—direct measurement of biomass—involves both satellite and ground measurements. Preliminary work with radar aboard aircraft and the space shuttle suggests that radar might be useful for direct determination of biomass (Hoffer et al. 1987; Kasischke et al. 1991), but much research has to be done to apply such an approach to stands of mixed age and mixed species where interpretation of the radar signal is complex. The European Space Agency has recently launched a satellite equipped with radar, so worldwide coverage is available if a reliable technique can be developed. The added advantage of radar over the other satellite data, mostly optical, is that it can penetrate cloud cover and, hence, is not limited.

Radar is unlikely ever to provide estimates of below-ground carbon stocks, including soils, however. Thus, direct field measurements of biomass will be necessary, both to calibrate radar, if it is to be used to provide in situ measurements, and to obtain estimates of soil carbon. For some areas of the world, maps of soil carbon already exist. Recording change in this pool, however, will require additional field measurements, with a frequency to be determined as a function of carbon per hectare and rate of deforestation.

For the purposes of monitoring carbon emissions, we should distinguish carbon stocks in the forest (live vegetation, dead plant material, soil organic matter) from carbon stocks removed from the forest (fuelwood, charcoal, pulpwood, sawlogs, and other products derived from wood). Identifying the change in land use allows indirect estimation of the fate of biomass and soil carbon. We have seen crude estimates of the amount of carbon lost under different types of land use in table 2.3. Thus, if land use can be determined from classification of satellite data or from ground studies, rough estimates of carbon emissions can be obtained. Although table 2.3 shows only the relative amounts of carbon lost, the rates of loss are reasonably well documented for a number of ecosystems and land uses (Houghton et al. 1987). Such estimates would be adequate in carrying out the first plan of a monitoring program, proposed above.

Still another way to estimate change in carbon stocks is through repetitive satellite monitoring and the use of a Geographic Information System. When satellite data is entered into a GIS, knowledge about any specific location on the ground accumulates. For example, information concerning the age of a clearing or recovering forest will be available from the data used to measure rates of deforestation and reforestation. The age of forests or cultivated lands, in turn, can be used to infer accumulation or release of carbon with independent data on rates of productivity, decay, and soil oxidation.

The biomass affected by logging can be determined in a similar way. Logged forests can be identified with high resolution Landsat TM data (30 m resolution) and, perhaps, with MSS data (80 m resolution). In general, the more selective the harvest (and the less disturbed the forest), the more difficult it is to observe it from space, but, at the same time, the less carbon is released. Nevertheless, if forests are monitored repeatedly, even the cumulative effect of selective logging will eventually show up.

The long-term net carbon flux is generally not sensitive to the proportion of original forest biomass going to products with different storage times. Products such as lumber last various lengths of time. Tracking wood products on the ground is a bookkeeping nightmare. On the other hand, not to track them takes away one of the incentives for reducing emissions of carbon: that is, the incentive to waste less wood, to burn less, and to put more into long-lasting structures.

I propose that the proportion of original biomass removed from the site during harvest and the distribution of this wood among products of different average lifetimes be approximated on the basis of currently available statistics on wood production. If countries are unsatisfied with these estimates and can demonstrate that they are putting more products into long-term storage, the estimates can be changed.

Degrading as opposed to deforesting land also reduces carbon stocks. We need first to establish a minimum definition of *disturbance*, in order to assess when land can be considered degraded. Monitoring with the higher resolution satellites (Landsat, SPOT) will detect areas affected by disturbance. Repeated monitoring will reveal whether the degradation is progressive, eventually leading to the replacement of the forest with some other ecosystem, or whether the forest returns in subsequent years. When degradation in forests is detected, ground checks or more intensive sampling should be used to determine the cause of the degradation and its extent.

The major difference between monitoring biomass burning and monitoring for deforestation or degradation lies in the frequency and resolution of the required coverage. The required resolution is lower, but monitoring for fires will have to be done much more frequently, even daily. The best candidate is AVHRR. In all cases, the initial biomass and its fate are important. Questions that need to be answered are: Is the burning related to deforestation or is it in a nonforest ecosystem, such as a grazed savanna? What is the frequency of burning? Is the frequency changing? Monitoring for burning should be integrated with monitoring for deforestation so that the type of ecosystems burned (forest or other) can be determined, and so that the correspondence between fires and deforestation can be evaluated.

The data requirements described above provide information on rates of deforestation and reforestation, fate of deforested land, land use, and amounts and fate of carbon stocks. These data can then provide estimates for emissions of carbon, methane, and nitrous oxide.

The approach used to calculate carbon emissions must account for all of the carbon present at the time of disturbance (Houghton et al. 1983, 1987; Houghton, Boone, et al. 1985; Detwiler and Hall 1988; Hall and Uhlig 1991). Some of the vegetation may be burned (released immediately to the atmosphere). The proportions released as $CO_2$, $CH_4$, and CO depend on the intensity of the burn. Some of the original vegetation may be removed in wood products. The average lifetimes of different wood products determine the course of the annual release of carbon over that lifetime. Dead plant material neither burned nor removed decays exponentially on site and releases carbon accordingly. The organic carbon content of the soils must also be accounted for. Soil carbon is generally lost with disturbance; the loss is related to type of land use. With reforestation, carbon accumulates on land in vegetation and soil, at rates that can be estimated from climatic and soil characteristics.

Emissions of greenhouse gases other than carbon dioxide, such as methane and nitrous oxide, can be quantified through biomass burning and land use. Methane is released from anaerobic environments, such as wetlands, rice paddies, and land grazed by ruminants. The quantities released can be estimated from crop productivity, cultivation period, stocking intensity, and diet of ruminants. Nitrous oxide is released from new pastures and fertilized agricultural lands. Thus, age of pasture, stocking rates, and fertilizer use are thought to be important in determining rates of emissions. At present, however, the quantitative relations between these factors and emission rates are only crudely known. Global emissions of methane and, particularly, nitrous oxide, are not well known.

Technical problems associated with data collection necessary for such calculations include cloud cover over much of the tropics, incomplete coverage of land area by current receiving stations, gaps of many years in the past for which no data exist, and inability to assess biomass or soil carbon using remote sensing technologies. Many of these technical difficulties could be remedied by an international monitoring program having the magnitude and intensity commensurate with the threat of climatic change.

Allowances can be set to incorporate cheating under the threshold of detection. How much deforestation can occur without being noticed? Probably very little, unless the deforestation is of many scattered clearings less than 0.1 ha in size. How much degradation can occur without being noticed? Even at the highest resolution (SPOT, 20 m on a side for multispectral data; 10 m on a side for panchromatic data), single trees can be harvested without detection. On the other hand, the harvest would become detectable if continued over years.

Some conceptual questions remain for the implementation of such a program. First, how should one account for natural variations in, for example, fires? Should wildfire emissions be counted as part of the country's quota, thus allowing regrowth in subsequent years to expand quotas? No. This program needs the fundamental recognition that humans must actively accommodate nature, rather than take advantage of good years and cry "unfair" in bad years. If the concentrations of greenhouse gases in the atmosphere are to be kept stable, human societies will have to adjust to nature's variability.

Second, what if forests lose carbon with warming? Will that count against forested countries and change their quotas? If forests were to become sources of carbon as a result of factors having nothing to do with local activities, they would still release less carbon than if they were cleared. Thus, the possibility that forests might leak carbon to the atmosphere should not be used to justify clearing them.

And finally, what if forests accumulate carbon from changes in climate or elevated concentrations of $CO_2$? Should such natural accumulations expand the country's quota? All three questions are related to changes outside of local control. Should such changes either relax or tighten the constraints agreed to prior to a full understanding of forest responses? The answer would seem to be "Yes." From the broadest perspective, the objective is to maintain the temperature of the earth within some range that can in principle be agreed upon. Fluctuations outside of that range may be caused by either human or natural processes. If the only control we have is over our own activities, however, then we must act within our power to compensate for the vagaries of nature. As noted earlier in this chapter, the management of forests from the perspective of the global carbon cycle or climate fails to acknowledge other benefits of forests. The temperature of the earth is only one of many criteria for planetary health.

The plan proposed here has not directly addressed the question of who would monitor forests or how checks and inspections would be carried out. The agency might be part of a global convention on climate and/or a global convention on forests. The agency would have to be international. The monitoring would have to be done with participation of all countries and with supervision from a supra-country council. An international agency with a charge analogous to that of the International Atomic Energy Agency seems appropriate.

**References**

Ajtay, G. L., P. Ketner, and P. Duvigneaud. 1979. "Terrestrial Primary Production and Phytomass." In *The Global Carbon Cycle*. SCOPE 13, ed. B. Bolin, E. T. Degens, S. Kempe, and P. Ketner. New York: John Wiley and Sons.

Brown, S., and A. E. Lugo. 1982. "The Storage and Production of Organic Matter in Tropical Forests and Their Role in the Global Carbon Cycle." *Biotropica* 14(3):161–187.

———. 1984. "Biomass of Tropical Forests: A New Estimate Based on Volumes." *Science* 223:1290–1293.

Brown, S., A. J. R. Gillespie, and A. E. Lugo. 1989. "Biomass Estimation Methods for Tropical Forests with Applications to Forest Inventory Data." *Forest Science* 35:881–902.

Cicerone, R. J., and R. S. Oremland. 1988. "Biogeochemical Aspects of Atmospheric Methane." *Global Biogeochemical Cycles* 2:299–327.

Cooper, C. F. 1982. "Carbon Storage in Managed Forests." *Canadian Journal of Forest Research* 13:155–166.

Crutzen, P. J., and M. O. Andreae. 1990. "Biomass Burning in the Tropics: Impact on Atmospheric Chemistry and Biogeochemical Cycles." *Science* 250:1669–1678.

Detwiler, R. P., and C. A. S. Hall. 1988. "Tropical Forests and the Global Carbon Cycle." *Science* 239:42–47.

Dregne, H. E. 1985. "Aridity and Land Degradation." *Environment* 27:16–20, 28–33.

FAO. 1990a. *Interim Report on Forest Resources Assessment 1990 Project*. Committee on Forestry, Tenth Session. Rome: Food and Agriculture Organization. COFO-90/8(a).

———. 1990b. *1989 Production Yearbook*. Rome: Food and Agriculture Organization.

————. 1991. "Second Interim Report on the State of Tropical Forests." Presented at the Tenth World Forestry Conference. Paris, September 1991.

FAO/UNEP. 1981. *Tropical Forest Resources Assessment Project.* Rome: Food and Agriculture Organization.

Fearnside, P. M. 1986. "Brazil's Amazon Forest and the Global Carbon Problem: Reply to Lugo and Brown." *Interciência* 11:58–64.

Fearnside, P. M., Tardin, A. T., and Filho, L. G. M. 1990. *Deforestation Rate in Brazilian Amazonia.* Brasilia: National Secretariat of Science and Technology.

Flint, E. P., and J. F. Richards. 1991. "Historical Analysis of Changes in Land Use and Carbon Stock of Vegetation in South and Southeast Asia." *Canadian Journal of Forest Research* 21:91–110.

Gorse, J. 1985. "Desertification in the Sahelian and Sudanian Zones of West Africa." *Unasylva* 37:2–18.

Grainger, A. 1987. "The Future Environment for Forest Management in Latin America." In *Management of the Forests in Tropical America: Prospects and Technologies,* ed. J. C. F. Colon, F. H. Wadsworth, and S. Branham, 1–9. Washington, D.C.: United States Department of Agriculture.

————. 1988. "Estimating Areas of Degraded Tropical Lands Requiring Replenishment of Forest Cover." *International Tree Crops Journal* 5:31–61.

Hall, C. A. S., and J. Uhlig. 1991. "Refining Estimates of Carbon Released from Tropical Land-Use Changes." *Canadian Journal of Forest Research* 21:118–131.

Hao, W. M., M. H. Liu, and P. J. Crutzen. 1990. "Estimates of Annual and Regional Releases of $CO_2$ and Other Trace Gases to the Atmosphere from Fires in the Tropics, Based on the FAO Statistics for the Period 1975–1980." In *Fire in the Tropical Biota,* ed. J. G. Goldhammer, 440–462. Berlin: Springer-Verlag.

Hoffer, R. M., D. F. Lozano-Garcia, and D. D. Gillespie. 1987. "Characterizing Forest Stands with Multi-Incidence Angle and Multi-Polarized SAR Data." *Advances in Space Research* 7(11): 309–312.

Houghton, J. T., G. J. Jenkins, and J. J. Ephraums, eds. 1990. *Climate Change: The IPCC Scientific Assessment.* Cambridge: Cambridge University Press.

Houghton, R. A. 1990a. "The Future Role of Tropical Forests in Affecting the Carbon Dioxide Concentration of the Atmoshphere." *Ambio* 19:204–209.

————. 1990b. "The Global Effects of Tropical Deforestation." *Environmental Science and Technology* 24:414–422.

————. 1991a. "Biomass Burning from the Perspective of the Global Carbon Cycle." In *Global Biomass Burning,* ed. J. S. Levine, 321–325. Cambridge, Mass.: MIT Press.

————. 1991b. "Releases of Carbon to the Atmosphere from Degradation of Forests in Tropical Asia." *Canadian Journal of Forest Research* 21:132–142.

————. 1991c. "Tropical Deforestation and Atmospheric Carbon Dioxide." *Climatic Change* 19:99–118.

Houghton, R. A., and D. L. Skole. 1990. "Carbon." In *The Earth as Transformed by Human Action*, ed. B. L. Turner, W. C. Clark, R. W. Kates, J. F. Richards, J. T. Mathews, and W. B. Meyer, 393–408. Cambridge: Cambridge University Press.

Houghton, R. A., J. D. Unruh, and P. A. Lefebvre. In Manuscript. "Current Land Use in the Tropics and Its Potential for Sequestering Carbon."

Houghton, R. A., J. E. Hobbie, J. M. Melillo, B. Moore, B. J. Peterson, G. R. Shaver, and G. M. Woodwell. 1983. "Changes in the Carbon Content of Terrestrial Biota and Soils Between 1860 and 1980: A Net Release of $CO_2$ to the Atmosphere." *Ecological Monographs* 53:235–262.

Houghton, R. A., R. D. Boone, J. M. Melillo, C. A. Palm, G. M. Woodwell, N. Myers, B. Moore, and D. L. Skole. 1985. "Net Flux of $CO_2$ from Tropical Forests in 1980." *Nature* 316:617–620.

Houghton, R. A., R. D. Boone, J. R. Fruci, J. E. Hobbie, J. M. Melillo, C. A. Palm, B. J. Peterson, G. R. Shaver, G. M. Woodwell, B. Moore, D. L. Skole, and N. Myers. 1987. "The Flux of Carbon from Terrestrial Ecosystems to the Atmosphere in 1980 Due to Changes in Land Use: Geographic Distribution of the Global Flux." *Tellus* 39B:122–139.

Houghton, R. A., W. H. Schlesinger, S. Brown, and J. F. Richards. 1985. "Carbon Dioxide Exchange between the Atmosphere and Terrestrial Ecosystems." In *Atmospheric Carbon Dioxide and the Global Carbon Cycle*, ed. J. R. Trablaka, 113–140. Washington, D.C.: Department of Energy. DOE/ER-0239.

Kasischke, E. S., L. L. Bourgeau-Chavez, N. L. Christenson, and M. C. Dobson. 1991. "The Relationship between Aboveground Biomass and Radar Backscatter as Observed in Airborne SAR Imagery." *Proceedings of the Third Airborne Synthetic Aperture Radar (AIRSAR) Workshop*, 11–21. Jet Propulsion Laboratory, Pasadena, California, 23–24 May 1991. JPL Pub. 91-30, August.

Kauppi, P. E., K. Mielikainen, and K. Kuusela. 1992. "Biomass and Carbon Budget of European Forests, 1971–1990." *Science* 256:70–74.

Lean, J., and D. A. Warrilow. 1989. "Simulation of the Regional Climatic Impact of Amazon Deforestation." *Nature* 342:411–413.

Lillesand, T. M., and R. W. Kiefer. 1979. *Remote Sensing and Image Interpretation*. New York: John Wiley and Sons.

Mabbutt, J. A. 1984. "A New Global Assessment of the Status and Trends of Desertification." *Environmental Conservation* 11:103–113.

Malingreau, J.-P., and C. J. Tucker. 1988. "Large-scale Deforestation in the Southeastern Amazon Basin of Brazil." *Ambio* 17:49–55.

Malingreau, J.-P., G. Stephens, and L. Fellows. 1985. "Remote Sensing of Forest Fires: Kalimantan and North Borneo in 1982–83." *Ambio* 14:314–321.

Melillo, J. M., J. R. Fruci, R. A. Houghton, B. Moore, and D. L. Skole. 1988. "Land-use Change in the Soviet Union between 1850 and 1980: Causes of a Net Release of $CO_2$ to the Atmosphere." *Tellus* 40B:116–128.

Myers, N. 1980. *Conversion of Tropical Moist Forests*. Washington, D.C.: National Academy Press.

———. 1984. *The Primary Source*. New York: W. W. Norton.

———. 1989. *Deforestation Rates in the Tropical Forests and Their Climatic Implications*. London: Friends of the Earth.

Olson, J. S., J. A. Watts, and L. J. Allison. 1983. *Carbon in Live Vegetation of Major World Ecosystems*. Washington, D.C.: United States Department of Energy. TR004.

Palm, C. A., R. A. Houghton, J. M. Melillo, and D. L. Skole. 1986. "Atmospheric Carbon Dioxide from Deforestation in Southeast Asia." *Biotropica* 18:177–188.

Post, W. M., W. R. Emanuel, P. J. Zinke, and A. G. Stangenberger. 1982. "Soil Carbon Pools and World Life Zones." *Nature* 298:156–159.

Ramanathan, V., L. Callis, R. Cess, R. Hansen, I. Isaksen, W. Kuhn, A. Lacis, F. Luther, J. Mahlman, R. Reck, and M. Schlesinger. 1987. "Climate-Chemical Interactions and Effects of Changing Atmospheric Trace Gases." *Reviews of Geophysics* 25:1441–1482.

Salati, E. 1987. "The Forest and the Hydrological Cycle." In *The Geophysiology of Amazonia: Vegetation and Climate Interactions*, ed. R. E. Dickinson, 273–296. New York: John Wiley and Sons.

Sanchez, P. A., and J. R. Benites. 1987. "Low-Input Cropping for Acid Soils of the Humid Tropics." *Science* 238:1521–1527.

Schlesinger, W. H.. 1984. "The World Carbon Pool in Soil Organic Matter: A Source of Atmospheric $CO_2$." In *The Role of Terrestrial Vegetation in the Global Carbon Cycle: Measurement by Remote Sensing*. SCOPE 23, ed. G. M. Woodwell. New York: John Wiley and Sons.

Setzer, A. W., and M. C. Pereira. 1991. "Amazonia Biomass Burnings in 1987 and an Estimate of their Tropospheric Emissions." *Ambio* 20:19–22.

Uhl, C., and R. Bushbacher. 1985. "A Disturbing Synergism between Cattle Ranch Burning Practices and Selective Tree Harvesting in the Eastern Amazon." *Biotropica* 17:265–268.

United States Bureau of the Census. 1987. *World Population Profile: 1987*. Washington, D.C.: United States Bureau of the Census.

Woods, P. 1989. "Effects of Logging, Drought, and Fire on the Structure and Composition of Tropical Forests in Sabah, Malaysia." *Biotropica* 21:290–298.

Woodwell, G. M., ed. 1984. *The Role of Terrestrial Vegetation in the Global Carbon Cycle: Measurement by Remote Sensing*. SCOPE 23. New York: John Wiley and Sons.

Woodwell, G. M., J. E. Hobbie, R. A. Houghton, J. M. Melillo, B. Moore, A. B. Park, B. J. Peterson, G. R. Shaver, and T. A. Stone. 1983. *Deforestation Measured by Landsat: Steps toward a Method*. Washington, D.C.: United States Department of Energy. TR005.

Woodwell, G. M., R. A. Houghton, T. A. Stone, R. F. Nelson, and W. Kovalick. 1987. "Deforestation in the Tropics: New Measurements in the Amazon Basin Using Landsat and NOAA Advanced Very High Resolution Radiometer Imagery." *Journal of Geophysical Research* 92:2157–2163.

Zachariah, K. C., and M. T. Vu. 1988. *World Population Projections*. Baltimore: Johns Hopkins University Press.

Zinke, P. J., A. G. Stangenberger, W. M. Post, W. R. Emanuel, and J. S. Olson. 1986. *Worldwide Organic Soil Carbon and Nitrogen Data*. Oak Ridge, Tenn.: Oak Ridge National Laboratory. ORNL/CDIC-18.

# 3

# Indigenous Knowledge in the Conservation and Use of World Forests

**Darrell A. Posey**

C onservation has always been a cultural question, although environmentalists have acted for decades as though the preservation of nature had nothing to do with the human species. Today, however, experienced environmentalists and conservationists recognize that unless people have a direct stake and interest in conservation, even the best-designed projects in the world stand little chance for long-term success. Meanwhile, scientists have begun to demonstrate how native peoples can teach us new models for sustained natural resource use and management.

Their ancient indigenous traditions, developed through millennia of experience, observation, and experiment, provide options for sustainable management of natural resources.

This chapter outlines the importance of traditional knowledge in the discovery of the tropical forests' biological riches—economic as well as ecological. My basic argument is that native peoples understand how to use and conserve the forest. Forests are currently being destroyed, in part, because of non-forest dwellers' lack of knowledge about how best to exploit their vast diversity of medicines, foods, natural fertilizers, and pesticides.

Fundamental to the preservation of the forest and their inhabitants is to show that the living forests are more valuable than the deforested land. The sad truth is that currently developers regard forest land as economically valuable for cattle, lumber, and gold—all of which are obtained only through the destruction of forests and savannas. Indigenous peoples can teach us how to give greater value to the living forest—but only if they survive, and we can learn to give them equal say in the future of this planet.

Accepting that native peoples, despite their different cultural habits, have much to teach us, but in their own language and with their own conceptual categories, seems one of the hardest lessons for modern society. Indeed, colonizing societies have historically used these differences to justify imposing their own version of order, progress, and civilization on native peoples, while exploiting those peoples' lands and natural resources. But as forests, particularly tropical forests, disappear, we are learning that their inhabitants may offer the best instruction in survival and forest use.

### Ethnobiology and Interdisciplinary Research

Ethnobiology combines anthropology with biology: it attempts to analyze what native peoples know about their environment. It is therefore a powerful science, because it joins the interdisciplinary and multidisciplinary forces of our Western cultural traditions to document, study, and give value to the knowledge systems of native peoples. Ethnobiologists present indigenous peoples not as exotic creatures with strange cultural habits but rather as societies living in close association with their environments. Most of them have done so for millennia. In Brazil, for example, archaeological evidence shows indigenous presence in South America for as many as 35,000 years.

Traditionally, Indians have been thought of as mere exploiters of their environments—not as conservers, manipulators, and managers of natural resources. Researchers are finding, however, that ecological systems in Amazonia hitherto presumed "natural" are, in fact, products of human manipulation (see, for example, Alcorn 1981, 1989; Anderson and Posey 1985; Balée 1989b; Balée and Gély 1989; Clement 1989; Denevan and Padoch 1987; Frickel 1959; Posey and Overal 1990). Likewise, extensive, ancient agricultural fallows reflect human-engineered genetic diversity (Balée 1989a; Denevan and Padoch 1987; Irvine 1989).

Because aboriginal populations were many times larger than today's indigenous populations, the extent of human influence can easily be underestimated. Ethnohistorical research with the Kayapó Indians, located in the Xingu River Basin in Pará, Brazil, for example (on whom I shall focus in this chapter), suggests that contact with European diseases came before face-to-face contact with the white man. Epidemics led to intragroup fighting and fission that, in turn, had interesting effects on species dispersal as well (Posey 1987). A form of nomadic agriculture developed for the cultivation of many domesticated and semidomesticated plant species in managed environments near trailsides, in abandoned villages, and at campsites (Posey 1986). Over millennia, the Kayapó spread seeds, tubers, and cuttings over vast areas of Brazil to insure the availability of valuable medicinal and edible species. Agricultural practices also spread, along with techniques to manage old fallows for wildlife and useful plants.

Over 76 percent of the plant species used by the Kayapó are not "domesticated." Nor can they be considered "wild," however, because they have been systematically selected for desirable traits and propagated in a variety of habitats. During times of warfare, the Kayapó could abandon their agricultural plots and survive on the semidomesticated species that for millennia had been scattered in spots known throughout the forest and savanna. Old agricultural field sites became hunting preserves and orchards, for the Indians had planted them when they were younger fields to mature for such purposes. In other words, agricultural areas were designed to develop into productive agroforestry plots dominated by semidomesticated species, thereby allowing the Kayapó to shift between being agriculturalists and hunter-gatherers. Such patterns appear to have been widespread in the lowland tropics and make archaic the traditional dichotomies of wild versus domesticated species, hunter-gatherers versus agriculturalists, and agriculture versus agroforestry.

Indigenous manipulation of environments in Amazonia was significant in molding other ecological systems. In the formation of "islands of forest" (Apêtê) in the campo-cerrado, for example, the Kayapó were found to have concentrated plant varieties collected from an area the size of Western Europe into a ten-hectare plot of Apêtê (Anderson and Posey 1985).

One of the major lessons learned from research in the 1980s is that apparently natural landscapes in Amazonia may be the products of diversified manipulation of ecological systems by vast aboriginal or even smaller ancient populations. Far too many biologists and ecologists, however, ignore the archaeological, historical, anthropological, and ethnobiological literature linking biodiversity with human activity. Research will now have to accommodate this information.

Indigenous knowledge comprises an integrated system of beliefs and practices distinctive to a cultural group. Among native peoples, in addition to information shared generally, there is specialized knowledge held by a few. Many indigenous groups have experts in soils, plants, animals, crops, medicines, and rituals. But each individual believes that he or she has the ability to survive alone in the forest. This offers great personal security and permeates the fabric of everyday life.

In an attempt to separate cultural interpretations by the investigator from explanations by the native, anthropologists and ethnobiologists adopted the linguistic distinction of *emic* and *etic*. Emic interpretations reflect cognitive and linguistic categories of the natives, whereas etic interpretations are those that have been developed by the researcher for purposes of analyses.

Sometimes the "consciousness" of knowledge is just a matter of putting an abstract label on a well-known, but unconscious, nonverbalized phenomenon. The native of any culture can become conscious of certain "commonsense" acts of "management" when alerted to them by the researcher. An Indian throwing seeds on the ground and stepping on them may discover that he or she, according to the researcher's interpretaton, has been managing an agricultural environment. "Yes, we do that," the Indian may say, "but that's not planting or management." At least the actions are agreed upon by researcher and researched, even if the emic and etic interpretations differ. But the informant too learns of the categories being used by the researcher and may even modify his or her own way of looking at his or her culture.

The distinction between "interpretation" and "reality" becomes even more complicated when dealing with higher levels of abstraction, such as spirits and mythological beings or forces. Native peoples generally make a point of saying that the forest for them is not just an inventory of natural resources but represents the spiritual and cosmic forces that make life what it is. What role, then, do metaphysical concepts play in management practices?

The Kayapó, for example, believe that old, abandoned village sites are full of spirits. Fear of spirits puts these old places off limits for many Indians. Only those who deal with spirits (shamans) or special hunting parties go there. Thus, these abandoned camps and villages effectively become protected reserves, with a high diversity of secondary growth that also attracts many animals. The spirits effectively serve as ecological protective agents.

Consider what the Kayapó call the *Pitu* plant (actually a grouping of plants from several botanical families). Pitu is one of the few plants that is thought to have a spirit—a very powerful spirit that once killed thousands of Indians. The spirit of this plant is so dangerous that if people touch it, or even go near it, they will die. It is said, however, that Pitu can be planted by shamans in secret, specialized medicinal gardens. Fear of coming into contact with Pitu is sufficient to keep out unwanted guests in these gardens and to guarantee the secrecy of the garden and its contents.

An etic interpretation of Pitu maintains that the fear of its spirit functions to protect medicinal gardens and restrict use to specialists of medicinal plants. It would be impossible, however, to find an Indian who would say: "Well, yes, the Pitu spirit functions in our society as an ecological management agent to protect our medicinal plantations."

So we return to the initial problem of interpreting the "reality" of native peoples. The problem of emic versus etic analyses has scared biologists and ecologists away from anthropology, which they consider unscientific. Botanists and zoologists do not, after all, have to confirm their scientific analyses with their biological subjects. Having only an etic level of analysis makes scientific investigation easier. When discussing human cultures, however, scientists must not confuse objectivity with the obscuring of reality. There is much to learn from the interpretation of indigenous myths, legends, and folk taxonomies, regardless of whether the methods meet the rigors of some scientific disciplines.

A debate rages within anthropology itself as to whether cultural interpretation can ever become "scientific." To the ethnobiologist, the debate seems rather inane. Ethnobiologists attempt to use all the scientific tools that can be borrowed from botany, zoology, geography, pedology, genetics, and ecology, among others. But that does not mean abandoning the quest for an emic view as well. If searching for native realities noses a bit too far into metaphysics or "fuzzy" science for some hardliners, then we must conclude that not everyone can or should be an ethnobiologist. We cannot forget, however, that attitudes regarding what is true science seriously divide the social and natural sciences.

One of the greatest barriers to interdisciplinary scientific investigation is the difference in research time frames used by social and natural scientists. Biologists consider a few months to be a reasonable field period, whereas anthropologists think in terms of years, in order to dominate a language well enough to delve into native perceptions of natural resources, concepts of management, mythological forces, and other levels of conscious or unconscious knowledge. This fundamental difference alone justifies the development of the hybrid field of ethnobiology, which trains students to weigh as equally important cognitive analyses of semantic fields and gathering basic biological and ecological data.

Another obstacle to research into traditional knowledge is the methodological problem of assessing the degree to which that knowledge is shared within a society. Even in the smallest societies, individuals do not know the same things. Scientists who have worked with native peoples have painfully learned this—or ignored it, to the detriment of their data analyses.

The Kayapó and other Jê peoples, for example, have highly specialized knowledge. Twenty-six percent of the approximately seven hundred people of the Kayapó village of Gorotire are medicinal curing specialists. Each specialist understands certain types of animal spirits that provoke diseases and can only be treated with a specific array of medicinal plants, magical songs, and curing rituals.

Roughly 15 percent of Gorotire inhabitants can identify and name at least thirty-five species of stingless bees (*Meliponidae*). The remaining 85 percent have difficulty recognizing more than eight. But some specialists can tick off fifty-four species, including details of their morphological characteristics, nesting habits, flight patterns, seasonal production of honey, and varied uses of their wax, bitumen, pollen, and honey.

To complicate matters, specialists frequently do not agree on details of knowledge. Two specialists in "fish diseases" (*tep kane*), for example, may vehemently disagree about which method for preparing "fish medicine plants" is most effective—or even which plants can be used for which type of the disease.

These methodological problems can be handled by trying to construct statistically significant survey and analytical methods to describe "typical" Kayapó knowledge, but such endeavors create nightmares for field researchers and result in questionable benefits. If the detailed knowledge of biological and ecological knowledge interests the researcher, then careful documentation, checking, and cross-checking to find anomalies and contradictions between informants will adequately advance ethnobiological research. More important, they will be enough to allow the researcher to advance hypotheses (see Posey 1986).

Most ethnobiological studies have tended to search in native knowledge only for what is already known from science. So we look for categories of plant use, animal behavior, ecological relations, soil types, and landscapes that already exist in our own knowledge system. Indigenous concepts may provide shortcuts or even breakthroughs in scientific investigation, however, that can then undergo the scientific treatment of hypothesis generation and testing. No ethnobiologist has ever insisted that traditional knowledge be taken at face value. Rather ethnobiologists hold that indigenous concepts should be used to guide researchers to look for unknown categories of knowledge or unknown ecological relations; in other words, it should offer researchers ways to generate new hypotheses with which to test indigenous concepts.

In this manner, ethnozoologists have "discovered" unknown species and subspecies of bees from the knowledge of tribal bee specialists and analyzed animal diets with the aid of skilled hunters, ethnopharmacologists have isolated active compounds in laboratories as a result of research with medicinal curers, ethnoecologists have carried out pioneering behavioral studies of little-known species with the help of native specialists, and ethnopedologists and ethnoagriculturalists have recognized soil-plant-animal complexes under the tutelage of veteran agriculturalists.

The decisions of scientists on how to propose their hypotheses based upon indigenous knowledge reveals the arbitrary nature of this basic step in the scientific pursuit, for the researcher frequently discards from her or his formulations the "unlikely" or "unbelievable" elements of informants' statements. But what constitutes the "unlikely" or "unbelievable" more often reflects the researcher's ability to grasp native reality than corresponds to any real scientific criteria. Nonetheless, proposing and testing hypotheses provides the methodological and theoretical bridge necessary to link scientific research with traditional knowledge.

### Kinds of Knowledge Available from Indigenous Peoples

A comprehensive overview of indigenous knowledge is, therefore, difficult (or even impossible) to represent because of its underlying cultural complexity. I can, however, identify some of the categories of our study of indigenous knowledge that indicate new research directions, even shortcuts, for Western science.

*Ethnoecology.* Ethnoecologists study native practices in local ecosytems. Indigenous peoples identify specific plants and animals as occurring within particular ecological zones. They have a well-developed understanding of animal behavior and also know which plants are associated with particular animals. Plant types, in turn, are associated with soil types. Each ecological zone represents a system of interactions among plants, animals, soils, and, of course, the people themselves.

The Kayapó, for example, recognize ecosystems that lie on a continuum between the poles of forest and savanna. They have identified a variety of types of savanna—savanna with few trees, savanna with many forest patches, savanna with scrub, and so on. But the Kayapó concentrate less on the differences among zones than on the similarities that cut across them. Marginal or open spots within the forest, for example, can have microenvironmental conditions similar to those in the savanna. The Kayapó take advantage of these similarities to exchange and spread useful species among zones, through transplanting seeds, cuttings, tubers, and saplings. Thus, they create considerable interchange among what we tend to see as distinctly different ecological systems.

Kayapó agriculture focuses upon the zones intermediate between forest and savanna, because there maximal biological diversity occurs. They often build villages in these transitional zones. The Kayapó not only recognize the richness of ecotones, the areas between ecological zones that form a transitional phase of growth, but they actually create them. They exploit secondary forest areas and create special concentrations of plants in forest fields, rock outcroppings, trailsides, and elsewhere.

The creation of forest islands, or Apêtê, demonstrates the extent to which the Kayapó can alter and manage ecosystems to increase biological diversity. Apêtê begin as small mounds of vegetation, about one to two meters around, and are created by transporting organic matter obtained from termite and ant nests to open areas in the field. Slight depressions are usually sought out because they are more likely to retain moisture. The Kayapó plant seeds or seedlings in these piles of organic material. The Apêtê are usually formed in August and September during the first rains of the wet season and then nurtured by the Indians as they pass along the savanna trails.

As Apêtê grow, they begin to look like upturned hats, with higher vegetation in the center and lower plants growing in the shaded borders. The Indians usually cut down the highest trees in the center to create a donut-hole center that lets the light into older Apêtê. Thus, a full-grown Apêtê has an architecture that creates zones that vary in shade, light, and humidity.

These islands become important sources of medicinal and edible plants, as well as places of rest. Palms, which have a variety of uses, figure prominently in Apêtê, as do shade trees. Even vines that produce drinkable water are transplanted here. Apêtê look so natural, however, that until recently scientists did not recognize that they were in fact human creations.

According to informants, of a total of 120 species inventoried in 10 Apêtê, about 75 percent could have been planted. Such ecological engineering requires detailed knowledge of soil fertility, microclimatic variations, and species' niches, as well as of the interrelations among the species that are introduced into these humanmade communities. The eating habits of deer and tapir are well known to the Indians, who stock forest islands with the animals' favorite foods. In this sense, Apêtê must be viewed as both agroforestry plots and hunting reserves.

The Kayapó are aware that some species develop more vigorously when planted together. They frequently speak of plants that are "good friends" or "good neighbors." One of the first of these "neighbor complexes" I was able to discover was the *tyruti-ombiqua*, or "banana neighbors." Among the plants that thrive near bananas are some of the *mekraketdja* ("child want not") plants, which are important in regulating fertility among the Kayapó (Elisabetsky and Posey 1989).

This diversity seems quite ordered to the Indian eye, with careful matchings between plant varieties and microenvironmental conditions. What appear to us as random field plantings turn out to have five more or less concentric zones, each with preferred varieties of cultivars and different cultivation strategies. Kayapó fields look like a real mess to Westerners used to nice "clean" plots with orderly, symmetrical rows.

*Ethnopedology.* In this discipline, scientists survey indigenous soil taxonomies, which show sophisticated horizontal and vertical distinctions based on texture, color, drainage qualities, friability, and stratification. Soil qualities are frequently related to indicator plant species that allow Indians to identify floral and faunal components associated with specific soil types, each of which is managed differently, according to individual characteristics. Sweet potatoes, for instance, prefer hotter soil and thrive in the center of fields where shade from the margins rarely penetrates. The plants must be well aerated, however, or soil compaction will smother the root system. Much hand work is necessary to turn over the soils, take out larger tubers, and replant smaller ones.

The Kayapó use various types of ground cover such as vegetation, logs, leaves, straw, and bark to affect moisture, shade, and temperature of local soils. Holes are sometimes filled with organic matter, refuse, and ash to produce highly concentrated pockets of rich soil. Old banana leaves, stalks, rice straw, and other organic matter are piled and sometimes burned in selected parts of fields to create additional local variations.

The Kayapó use the ash of dozens of types of plants, each said to have certain qualities preferred by specific cultivars. The ash is usually prepared from the vines, shucks, stalks, and leaves of plants that have been cut or uprooted during harvesting or weeding. Sometimes piles of organic matter are made, with the different varieties carefully separated and allowed to dry in the sun until they will give a complete burn. The ashes are then distributed to the appropriate part of the field.

*Ethnozoology.* With ethnozoology, scientists have expanded their knowledge of animal species and behavior. Like other indigenous groups, the Kayapó conscientiously study animal anatomy, giving special attention to stomach contents of game animals, which tells them what plants will attract the animals and thus make them easier to hunt. They are also astute observers of many aspects of animal behavior. The Kayapó encourage their children to learn the behavior patterns and feeding habits of different animal species, which are considered to have their own personalities. Part of this knowledge is gained through rearing pets. In a survey done with Kent Redford I found over sixty species of birds, reptiles, snakes, amphibians, mammals—even spiders—being raised in the village.

Kayapó use a precise knowledge of insect behavior to control agricultural pests. For example, nests of "smelly ants"—*mrum kudja*, of the genus *Azteca*—are deliberately placed by the Indians in gardens and on fruit trees that are infested with leaf-cutting ants (*Atta* spp.). The pheromones of the smelly ants repel the leaf-cutters. These protective ants are also highly prized for their medicinal properties. They are frequently crushed, and their highly aromatic scents inhaled to open up the sinuses.

The Indians cultivate several plants containing extra-floral nectars, often on the leaves or stems, which attract predatory ants to serve as bodyguards for the plant. They plant banana trees to form a living wall around their fields, because predatory wasps prefer to nest under the leaves.

Stingless bees (*Meliponidae*) are one of the most valued insect resources. During the dry season, groups of men go off for days to find honey, which they often drink at the collection site. Beeswax is brought back to the village, to be burnt in ceremonies and used in many artifacts. One of my knowledgeable and patient teachers, the shaman Kwyrà-kà, was a great expert on stingless bees. When I went with him and his son, Ira, up river to hunt, we spent most of our time searching for honey. His son had learned to draw at the missionary school and loved to sketch the bees' nests. Having been trained originally in entomology, I realized what a gold mine of information these two Indians possessed about the behavior of what our scientists still consider little-known species.

*Ethnomedicine and Ethnopharmacology.* These interdependent disciplines explore indigenous medical and pharmacological knowledge. Almost every Kayapó household has its complement of common medicinal plants, many of which are domesticates or semidomesticates. Shamans specialize in the treatment of particular diseases. Diarrhea and dysentery remain the major killers in the humid tropics, where the Kayapó have classified over 150 types of diarrhea and dysentery, each of which is treated with its own specific medicine. Folk categories can be more elaborate and detailed than their "civilized" counterparts. Ethnopharmacologists and physicians frequently forget that disease categories, like all intellectually perceived phenomena, are culturally classified rather than universal.

Kayapó plant classification is based on each plant's pharmacological properties—that is, for which diseases it can serve as a cure. The shaman Beptopoop was the first Kayapó to show me how rare medicinal plants could be brought from distant areas and transplanted to places near home or in special medicinal rock gardens. He specializes in curing the bites and stings of snakes, lizards, and scorpions, and knows the most minute details of their behavior. I got a feeling for the sophistication of Kayapó plant knowledge when he showed me how to graft a prized species for treating scorpion sting onto more common stock that grew near his favorite forest trail. Indian plant categories cut across morphologically based botanical groupings. Nevertheless, these taxonomies often correlate closely to our botanical classification.

In addition to the discovery of medicinal plants, the larger branch of *ethnobotany* can establish new uses for known species and document the uses of unknown ones. *Kupa (Cissis gonglyodes)*, for instance, is an edible domesticate known only to the Kayapó and some of their relatives. An estimated 250 plants have been collected that are used for their fruits alone.

*Ethnoagriculture and Agroforestry.* These studies may offer one of the best hopes for securing the future of forests and, consequently, of forest dwellers themselves. Indigenous agriculture begins with a forest opening into which useful species are introduced and ends with a mature forest of concentrated resources, including game animals. The cycle is repeated when the "old-field" forests develop canopies too high and dense for efficient production and are cleared again.

Agricultural plots are designed to be productive throughout this reforestation cycle. Contrary to persistent beliefs about indigenous slash-burn agriculture, fields are *not* abandoned a few years after initial clearing and planting. On the contrary, old fields offer an important concentration of diverse resources long after primary cultivars have disappeared. Kayapó "new fields," for example, peak in production of principal domesticated crops in two or three years but continue to bear produce for many years after: sweet potatoes return for four to five years, yams and taro for five to six years, and papaya and banana for five or more years. The Kayapó revisit old fields seeking these lingering riches.

Fields take on new life as plants in the natural reforestation sequence begin to appear. These plants soon constitute a type of forest for which the Kayapó have a special name that means "mature old fields." Such fields provide a wide range of useful products and are especially valuable for their concentrations of medicinal plants. Old fields also attract wildlife to their abundant, low, and leafy plants. Intentional dispersal of old fields and systematic hunting extends human influence over the forest by providing, in effect, large game farms near centers of human population.

The Kayapó do not make a clear distinction between fields and forest, nor between wild and domesticated species. Collected plants are transplanted into concentrated spots near trails and campsites to produce "forest fields." The trailsides themselves become planting zones. It is not uncommon to find trails composed of four-meter-wide cleared strips of forest.

The processes of species domestication, frequently assumed to be complete, still occupy indigenous groups like the Kayapó. With the team members of the Kayapó Project, we have collected literally hundreds of plant varieties that have been systematically selected by the Kayapó and planted in ecological systems modified by humans. It is fair to conclude that similar activities have gone on and continue to go on throughout the Amazon among native peoples. Thus, plant species are probably being led toward domestication as you read this book.

### Knowledge as Business

Industry and business discovered many years ago that indigenous knowledge means money. In the earliest days of colonialism, extractive products formed the basis for colonial wealth, accruing to the colonizers. Later, pharmaceutical industries became, and remain today, the major exploiters of traditional medicinal knowledge for major products and profits.

The annual world market value for medicines derived from plants discovered through indigenous peoples is $43 billion. Estimated sales for 1989 of three major natural products in the United States alone were: Digitalis, $85 million; Resperine, $42 million; Pilocarpine, $28 million (E. Elisabetsky, personal communication). Although no comparable figures are published for natural insecticides, insect repellents, and genetic plant materials acquired from native peoples, the annual potential for such products is easily equal to that of medicinal plants. Research into these products is only beginning, with projections of their market values exceeding those of all other food and medicinal products combined. The international seed industry, much of which derived original genetic materials from crop varieties "selected, nurtured, improved and developed by innovative Third World farmers for hundreds, even thousands of years" (Hurtado 1989), accounts for over $15 billion per year alone.

Likewise, natural fragrances, dyes, and body and hair products have become major world markets. Figures from the Body Shop, considered to be one of the successes in international enterprise, show annual sales of $90 million with a growth rate last year of 60 percent (*Time* 1990). The three hundred Body Shop products are derived from plants, are not tested on animals, and mostly come from "Third World" countries. These products are marketed as coming from ecologically sustainable projects managed by the native peoples themselves. The success of

Anita Roddick, founder of the British company, earned her the title of Britain's Retailer of the Year in 1990. Such renown does not go unnoticed by the hundreds of would-be imitators of her marketing strategy.

Growing interest and catapulting markets in "natural" food, and medicinal, agricultural, and body products require increased research into traditional knowledge systems. Now more than ever the intellectual property rights of native peoples must be protected and just compensation for their knowledge guaranteed. We cannot simply rely upon the goodwill of companies and institutions to do right by indigenous peoples. If something is not done now, mining the riches of indigenous knowledge will become the latest—and ultimate—form of neo-colonial exploitation of native peoples.

Ecologists are justifiably concerned about the ecological impact when natural products become too successful. Developers of these products tend to create monocultures of their cash crops. Many scientists worry that international demands may spell the end of biodiversity rather than encourage conservation of natural resources, as initially desired. Michael Soulé and Kathryn Kohm (1989) outline this concern:

> Increased pressure on biological resources arises because of increasing human populations, changing consumption patterns, and new technologies. Although agricultural intensification will continue to be necessary, its impact on biological resources is not predetermined. Conservation poses important research questions relevant to the design of new production technologies and land use systems: Can biologically diverse and low energy technologies be extended and/or intensified? Can production systems be differentially intensified so as to maintain biological diversity in other parts of a system? How does increased exploitation of specific species affect other species and general system properties?

Provoking cultural change can be equally disconcerting. By establishing mechanisms for monetary compensation of native peoples, are we not also establishing the means to destroy their societies, through subverting their cultures to materialism and consumerism? Given current realities, such concerns offer only romantic worries. The fact is that indigenous societies and their natural environments are *already* being destroyed by the dramatic expansion of industrialized society. Pharmaceutical companies and natural-products companies have tasted success in their efforts; they will not go away.

Certainly, deciding what represents just compensation and how such benefits should be distributed opens a Pandora's box. But not to open this box is to accept the ethical and moral responsibility of paternalism (we from "advanced societies" know what is good for the native because we have already made the mistakes of squandering our cultural and natural wealth) that has undermined indigenous independence since the first wave of colonialism. Native peoples must have the right to chose their own futures. Without economic independence, such a choice is not possible.

### Suggestions for Future Research

I offer here a few suggestions to guide future research into traditional natural resource management techniques and their applications. First, we should study genetic manipulation of flora and fauna by native peoples, which remains relatively little known. Systematic research is needed into selection procedures, decisions behind choices for different species and varieties, inter- and intratribal variations in selection of variables, and evolutionary consequences of differential selective behavior.

More extensive studies should be made to describe how native peoples modify landscapes and environments. Fire is immensely important in almost all traditional management systems, yet few details of fire use are available (when to burn, what can be burned and when, temperature of the burn, frequency of burning, protection from burning, products of burning—for example, charcoal, ashes, stumps, and charred root systems). Likewise, little is known of the effects of fire on plant communities (which species are destroyed, which are stimulated, effects of burning on blooming and fruiting timing and production, and modification of morphological structures as a result of burning).

Most research to date has focused on indigenous manipulation of forests and agricultural plots, yet evidence shows that other habitats may also be significantly modified. Scrub forests and savannas, for example, are certainly molded by the use of fire. Creation of Apêtê in campo-cerrado by the Kayapó shows that ecological communities can be humanmade, with a high diversity of useful plants from different areas. The Kayapó also modify hillsides, trails, and even rocky outcroppings to maximize resource availability (see Posey 1985). To get a more complete picture of indigenous adaptation and management, future research should emphasize investigations into these less-studied habitats.

Rivers, streams, and seashores can also be modified by indigenous activities. Although extensive documentation of the dependence of native peoples on fresh and salt water species exists, only scattered evidence (see Chernela 1989) is available on how native peoples manage these important resources.

Baseline studies are necessary to establish wildlife population composition, numbers, and carrying capacities. Many researchers have observed that old fields are favored hunting areas for Indians. I believe that where old fallow management is practiced, wildlife populations may be higher in number and species diversity. This hypothesis goes against traditional thinking, which insists upon the innate destructiveness of Indians to wildlife populations. Baseline data from uninhabited areas must be available for comparison with managed, inhabited areas, therefore.

Ethnolinguistic studies can provide us with valuable data on the historic relationships between peoples and the botanical interchanges that occurred between them, or the cognitive geographical maps that link resources with the physical world. Myths, legends, ceremonies, rituals, and songs are filled with ecological and biological information, but very few studies have systematically analyzed their content.

More long-term sequence studies of fallows and other managed habitats are needed to understand the ecological transitions that accompany growth. These studies, coupled with knowledge of the natives' use of resources in the growth sequences, can give a much clearer picture of options for forestation and reforestation schemes.

Multidisciplinary research teams should be encouraged, despite the conceptual difficulties and those of timeframes that separate disciplines. Focusing upon the study of traditional knowledge can, indeed, be used to reunite the fragmented disciplines of science. Special ethnobiological academic and training programs should be established to develop the interdisciplinary field, combining the methods and techniques of ecology, biology, anthropology, and linguistics. Ethnobiological research centers should also be created to coordinate multidisciplinary research and design programs for the application of traditional knowledge. These centers should include laboratories for the analyses of medicinal compounds, nutritional value of edible plants, and chemical characteristics of natural fertilizers, among others.

Little of this can be accomplished unless major international policy shifts by governments, lending institutions, research institutions, private foundations, and industry support research into the use of traditional knowledge and its potential application in solving modern world problems. Applied research centers need to be established so that experimental plots, analytical laboratories, and field stations can investigate ways for sustained resource management based on indigenous models. Businesses can develop new categories of the use of plant and animal products (cosmetics, alternative building materials, and other natural products), creating international markets for these products. Once the diversity of native products and their market potentials are known, it will be possible to design reforestation and forestation programs and forest reconstruction projects that include native peoples as *intellectual* participants in all stages of project planning and implementation. It is time to develop alternative market infrastructures that are sensitive to indigenous peoples and their needs. If market mechanisms cannot be created that respect both the ecological balance and the intellectual integrity of indigenous peoples, then there is little hope that our consumer-based society can survive.

### References

Alcorn, J. B. 1981. "Huastec Noncrop Resource Management: Implications for Prehistoric Rain Forest Management." *Human Ecology* 9:395–417.

———. 1984. *Huastec Mayan Ethnobotany.* Austin: University of Texas Press.

———. 1989. "Process as Resource: The Traditional Agricultural Ideology of Bora and Huastic Resource Management and Its Implications for Research." In *Resource Management in Amazonia: Indigenous and Folk Strategies,* ed. D. A. Posey and W. L. Balée, 63–77. Advances in Economic Botany 7. New York: New York Botanical Gardens.

Anderson, A., and D. A. Posey. 1985. "Manejo de Cerrado Pelos Indios Kayapó." *Boletim do Museu Paraense Emilio Goeldi: Botânica* 2(1):77–98.

Balée, W. L. 1989a. "Cultura na Vegetação da Amazônia." In *Biologia e Humana da Amazônia,* ed. W. A. Neves, 105–109. Belém, Brazil: Museu Paraense Emilio Goeldi.

———. 1989b. "The Culture of Amazonian Forests." In *Resource Management in Amazonia: Indigenous and Folk Strategies,* ed. D. A. Posey and W. L. Balée, 1–21. Advances in Economic Botany 7. New York: New York Botanical Gardens.

Balée, W. L., and A. Gély. 1989. "Managed Forest Succession in Amazonia: The Ka'apor Case." In *Resource Magagement in Amazonia: Indigenous and Folk Strategies,* ed. D. A. Posey and W. L. Balée, 129–148. Advances in Economic Botany 7. New York: New York Botanical Gardens.

Boster, J. S. 1984. "Classification, Cultivation, and Selection of Aguaruna Cultivars of Manihot esculenta (Euphorbiaceae)." In *Ethnobotany in the Neo-Tropics,* ed. G. T. Prance and J. A. Kallunki, 34–47. Advances in Economic Botany 1. New York: New York Botanical Gardens.

Carneiro, R. 1978. "The Knowledge and Use of Rain Forest Trees by the Kuikuru Indians of Central Brazil." In *The Nature and Status of Ethnobotany*, ed. R. Ford, 201–216. Anthropological Papers No. 67. Ann Arbor: University of Michigan Museum of Anthropolgy.

Chernela, J. 1989. "Managing Rivers of Hunger: The Tukano of Brazil." In *Resource Management in Amazonia: Indigenous and Folk Strategies*, ed. D. A. Posey and W. L. Balée, 238–248. Advances in Economic Botany 7. New York: New York Botanical Gardens.

Clements, C. R. 1989. "A Center of Crop Genetic Diversity in Western Amazonia: A New Hypothesis of Indigenous Fruit-Crop Distribution." *Bioscience* 39(9):624–630.

Conklin, H. C. 1957. "Hanunoo Agriculture: On an Integral System of Shifting Cultivation in the Philippines." FAO Forestry Development Paper 12. Rome: Food and Agriculture Organization.

Denevan, W. M., and C. Padoch. 1987. *Swidden-Fallow Agroforestry in the Peruvian Amazon*. Advances in Economic Botany 5. New York: New York Botanical Garden.

Elisabetsky, E., and D. A. Posey. 1989. "Use of Anticonceptual and Related Plants by the Kayapó Indians (Brazil)." *Journal of Ethnopharmacology* 26(3): 299–316.

Frikel, P. 1959. "Agricultura dos Indios Mundurucú." *Boletim do Museu Paraense Emilio Goeldi: Antroplogia* 8:1–41.

Hecht, S. B., and D. A. Posey. 1989. "Preliminary Results on Soil Management Techniques of the Kayapó Indians." In *Resource Magagement in Amazonia: Indigenous and Folk Strategies*, ed. D. A. Posey and W. L. Balée, 174–188. Advances in Economic Botany 7. New York: New York Botanical Gardens.

Hurtado, M. E. 1989. "Seeds of Discontent." *South*, September, 95–96.

Irvine, D. 1989. "Succession Management and Resource Distribution in an Amazonian Rain Forest." In *Resource Magagement in Amazonia: Indigenous and Folk Strategies*, ed. D. A. Posey and W. L. Balée, 223–237. Advances in Economic Botany 7. New York: New York Botanical Gardens.

Johnson, A. 1989. "How the Machiguenga Manage Resources: Conservation or Exploitation of Nature?" In *Resource Management in Amazonia: Indigenous and Folk Strategies*, ed. D. A. Posey and W. L. Balée, 213–222. Advances in Economic Botany 7. New York: New York Botanical Gardens.

Posey, D. A. 1983. "Indigenous Knowledge and Development: An Ideological Bridge to the Future." *Ciência e Cultura* 35(7):877–894.

———. 1985. "Indigenous Management of Tropical Forest Ecosystems: The Case of the Kayapó Indians of the Brazilian Amazon." *Agroforestry Systems* 3(2):139–158.

———. 1986. "Topics and Issues in Ethnoentomology with Some Suggestions for the Development of Hypothesis-Generation and Testing in Ethnobiology." *Journal of Ethnobiology* 6(1):99–120.

Posey, D. A., and W. L. Overal. 1990. *Ethnobiology—Implications and Applications.* Proceedings of the First International Congress of Ethnobiology. Belém, Brazil: Museu Paraense Emilio Goeldi. CNPq.

Ribeiro, B. G., and Tolamån Kenhiri. 1989. "Rainy Seasons and Constellations: The Desana Economic Calendar." In *Resource Magagement in Amazonia: Indigenous and Folk Strategies,* ed. D. A. Posey and W. L. Balée, 97–114. Advances in Economic Botany 7. New York: New York Botanical Gardens.

Salick, J. 1989. "Ecological Basis of Amuesha Agriculture, Peruvian Upper Amazon." In *Resource Magagement in Amazonia: Indigenous and Folk Strategies,* ed. D. A. Posey and W. L. Balée, 189–212. Advances in Economic Botany 7. New York: New York Botanical Gardens.

Soulé, M. E., and Kohm, K. A. 1989. *Research Priorities for Conservation Biology.* Washington, D.C.: Island Press.

*Time.* 1990. "Enterprising Ecologists." 23 April, p. 82.

# 4

# From Empty-World Economics to Full-World Economics: A Historical Turning Point in Economic Development

**Herman E. Daly**

Threvolution of the human economy has passed from an era in which manmade capital represented the limiting factor in economic development (an "empty" world) to an era in which increasingly scarce natural capital has taken its place (a "full" world). Economic logic tells us that we should maximize the productivity of the scarcest (limiting) factor, as well as try to increase its supply. This means that economic policy should be

The original version of this chapter appeared as "From Empty-World Economics to Full-World Economics: Recognizing an Historical Turning Point in Economic Development," in *Environmentally Sustainable Economic Development: Building on Brundtland*, edited by Robert Goodland, Herman Daly, Salah El Serafy, and Bern Van Droste. Copyright © UNESCO 1991. Reprinted by permission.

designed to increase the productivity and amount of natural capital rather than the productivity and accumulation of manmade capital, as was appropriate in the past, when manmade capital constituted the limiting factor. But few until now have recognized that we have not only reached an economic turning point, we have passed it.

### Reasons the Turning Point Has Not Been Noticed

Why has this transformation from a world relatively empty of human beings and their capital to a world relatively full of these not been noticed by economists? If such a fundamental change in the pattern of scarcity is real, as I think it is, then how could it be overlooked by economists, whose job it is to pay attention to the pattern of scarcity? Some economists (Boulding 1964; Georgescu-Roegen 1971) have indeed signalled the change, but their voices have been largely unheeded.

One reason is the deceptive acceleration of exponential growth. With a constant rate of growth the world will go from half full to full in one doubling period—the same amount of time that it took us to go from 1 to 2 percent full. And the doubling time itself has shortened, compounding the deceptive acceleration. If we take the percentage of the net product of land-based photosynthesis appropriated by humans as an index of how full the world is of humans and their furniture, then we can say that it is 40 percent full. This means that we use, directly and indirectly, about 40 percent of the net primary product of land-based photosynthesis (Vitousek 1986) Taking thirty-five years as the doubling time of the human scale (population times per-capita resource use) and calculating backward, we find that we reached the present 40 percent from only 10 percent full in just two doubling times, or seventy years—an average lifetime. Also *full* here means 100 percent human appropriation of the net product of photosynthesis, which on the face of it would seem to be ecologically unlikely and socially undesirable. (In a full world, only the most recalcitrant species would remain wild—all others would be managed for human benefit). In other words, effective fullness occurs at less than 100 percent human preemption of net photosynthetic product, and there is evidence that human carrying capacity is reached at less than even the existing 40 percent (Goodland 1991). So, the world has rapidly gone from relatively empty to relatively full. Although 40 percent may technically be less than half, we had better think of it as indicating relative fullness, because it is only one doubling time away from 80 percent, a figure which represents excessive fullness.

This change has exceeded the speed with which fundamental economic paradigms usually shift. According to physicist Max Planck, a new scientific paradigm triumphs not by convincing the majority of its opponents, but because its opponents eventually die. There has not yet been time for the empty-world economists to die—and meanwhile they have been cloning themselves faster than they are dying, by maintaining tight control over their guild. The disciplinary structure of knowledge in modern economics is far tighter than that of turn-of-the-century physics, Planck's model. Full-world economics is not yet accepted as academically legitimate; indeed it is not even recognized as an academic challenge (Daly and Cobb 1989).

Another reason few have noted the watershed change in the pattern of scarcity is that in order to speak of a *limiting* factor, factors must be thought of as complementary rather than substitutable. If factors can substitute for each other, then a shortage of one does not significantly limit the productivity of the other. A standard assumption of neoclassical economics has been that factors of production are highly substitutable. Although other models of production deny the substitutability of factors (consider the Leontief model, which is wholly complementary), it remains the dominant economic assumption. Consequently, economists have pushed the very idea of a limiting factor into the background. If factors are substitutes rather than complements, there can be no limiting factor and hence no new era based on a change of the limiting role from one factor to another. We must, therefore, be very clear on the issue of complementarity versus substitutability (Takayama 1985).

Productivity of manmade capital is more and more limited by the decreasing supply of complementary natural capital. In the past, when the scale of the human presence in the biosphere was low, manmade capital played the limiting role. The switch from manmade to natural capital as the limiting factor is thus a function of the increasing scale and impact of the human presence. Natural capital is the stock that yields the flow of natural resources—the forest that yields the flow of cut timber; the petroleum deposits that yield the flow of pumped crude oil, the fish populations in the sea that yield the flow of caught fish. The complementarity of natural and manmade capital should be clear when we ask, What good is a sawmill without a forest? a refinery without petroleum deposits? a fishing boat without populations of fish? Beyond some point in the accumulation of manmade capital the limiting factor on production must become remaining natural capital. The limiting factor determining the fish catch, for example, is the reproductive capacity of fish populations, not the number of fishing boats; for gasoline, petroleum deposits, not refinery capacity; and for many types of wood, the amount of forest remaining, not sawmill capacity. Already, Costa Rica and Peninsular Malaysia, among others, now must import logs to keep their sawmills employed. A country can accumulate manmade capital and deplete natural capital to a greater extent only if another country does it to a lesser extent—Costa Rica must import logs from somewhere. The demands of complementarity between manmade and natural capital can be evaded within a nation only if they are respected between nations.

Of course, simply multiplying specific examples of complementarity between natural and manmade capital will never suffice to prove my general case. But the examples given above at least serve to add concreteness to more general arguments. Because of the complementary relation between manmade and natural capital, the very accumulation of manmade capital puts pressure on natural capital stocks to supply an increasing flow of material. When that flow reaches a size that can no longer be maintained, exploiters are tempted to supply the annual flow unsustainably by liquidating natural capital stocks, thus postponing the collapse in value of their complementary manmade capital. Indeed, in the era of empty-world economics, natural resources and natural capital were considered free goods (except for extraction or harvest costs). Consequently, the value of manmade capital was under no threat from scarcity of a complementary resource. In the era of full-world economics, however, this threat is real and can be met by liquidating stocks of natural capital, which will temporarily keep up the flows of natural resources that support the value of manmade capital. Hence the problem of sustainability.

### Complementarity versus Substitutability

The next question must be, Can we replace natural resources with manmade capital? Clearly one resource can substitute for another—we can transform aluminum instead of copper into electric wire. We can also substitute labor for capital or capital for labor to a significant degree, even though the characteristic of complementarity limits this in important ways. For example, we can have fewer carpenters and more power saws or fewer power saws and more carpenters and still build the same house. But more pilots cannot substitute for fewer airplanes, once the airplanes are fully employed. In other words, one resource can substitute for another, albeit imperfectly, only if both play the same qualitative role in production—both are raw materials undergoing transformation into a product. Likewise, capital and labor can be substitutable to a significant degree because both play the role of agent of transformation of resource inputs into product outputs. If we wish to substitute across the roles of transforming agent and material undergoing transformation (efficient cause and material cause), however, the possibilities become limited, for the characteristic of complementarity dominates. After all, we cannot make the same house with one amount of lumber that we make with half that amount, no matter how many extra power saws or carpenters we try to substitute. Of course, we might substitute brick for lumber, but then we face the analogous limitation—we cannot substitute masons and trowels for bricks.

Natural capital (natural resources), therefore, complements rather than substitutes for manmade capital. The neoclassical assumption of a near-perfect substitutability between natural resources and manmade capital seriously distorts reality, the excuse of "analytical convenience" notwithstanding. To see this, just imagine that manmade capital were in fact a perfect substitute for natural resources. Then it must also be the case that natural resources make a perfect substitute for manmade capital. Yet if that were so, we would have had no reason to accumulate manmade capital, being already endowed by nature with perfect substitutes! Historically, of course, we accumulated manmade capital long before natural capital was depleted, precisely because we needed manmade capital to make effective use of the natural capital (complementarity!). It is amazing that the substitutability dogma should be held with such tenacity in the face of such an easy reductio ad absurdum. Add to this the fact that capital itself requires natural resources for its production—the substitute itself requires the input being substituted for—and it becomes quite clear that manmade capital and natural resources are fundamentally complements, not substitutes. Substitutability of capital for resources is limited to reducing waste of materials, for example, collecting sawdust and using a press (capital) to make particle board. But no amount of substitution can ever reduce the mass of material resource inputs below the mass of the outputs, given the law of conservation of matter and energy.

Substitutability of capital for resources in aggregate production functions reflects largely a change in the total product mix from resource-intensive to different capital-intensive products. It is an artifact of product aggregation, not factor substitution (that is, along a given product isoquant). It is important to emphasize that I am attacking the latter meaning of substitution here—in other words, producing a given physical product with fewer natural resources and more capital. No one denies that we can produce a different product or a different product mix with fewer resources. Indeed, new products frequently provide the same or better service while using fewer resources, and sometimes less labor and less manmade capital as well. Such production constitutes technical improvement, not substitution of capital for resources. Light bulbs that give more lumens per watt represent technical progress, qualitative improvement in the technology, not the substitution of a quantity of capital for a quantity of natural resource in the production of a given quantity of a product.

Perhaps economists who claim that capital is a near-perfect substitute for natural resources are speaking loosely and metaphorically. Perhaps they are counting as "capital" all improvements in knowledge, technology, managerial skill, and the like—in short, anything that increases the efficiency with which resources are used. Under this usage, capital and resources would by definition be substitutes in the same sense that more efficient use of a resource is a substitute for using more of the resource. But to define capital as efficiency would make a mockery of the neoclassical theory of production, where efficiency is a ratio of output to input, and capital is a quantity of input.

For earlier economists, however, human presence in the biosphere had not yet threatened the production of manmade capital. Now, the productivity of manmade capital is more and more limited by the decreasing supply of complementary natural capital. This is a function of the increasing intrusion of human enterprise into the biosphere.

### What Is Natural Capital?

Thinking of the natural environment as "natural capital" can be useful within limits. We may define *capital* broadly as "a stock of something that yields a flow of useful goods or services." Traditionally, *capital* was defined as "produced means of production," which I am here calling manmade capital, in distinction to natural capital. Although not made by man, natural capital nevertheless functions as a stock that yields a flow of useful goods and services. We can distinguish renewable from nonrenewable and marketed from nonmarketed natural capital, giving us four cross categories. Pricing natural capital, especially nonmarketable natural capital, so far offers an intractable problem, but one that need not be faced here. All that need be recognized for the argument at hand is that natural capital consists of physical stocks that are complementary to manmade capital. We have learned to use a concept of "human capital," a body of human knowledge and skills, which departs even more fundamentally from the standard definition of capital. Human capital cannot be bought and sold, although it can be rented. Although it can be accumulated, it cannot be inherited as can ordinary manmade capital, but must be learned anew by each generation. Natural capital, however, resembles traditional manmade capital in that it can be bequeathed. Overall, the concept of natural capital departs less from the traditional definition of capital than does the commonly used notion of human capital.

A troublesome subcategory of marketed natural capital, intermediate between natural and manmade, might be called *cultivated natural capital*, consisting of such things as plantation forests, herds of livestock, agricultural crops, fish bred in ponds, and others. Cultivated natural capital supplies the raw material input complementary to manmade capital, but does not provide the wide range of natural ecological services characteristic of natural capital proper. Eucalyptus plantations, for example, supply timber to the sawmill and may even reduce erosion but do not provide a wildlife habitat nor preserve biodiversity. Investment in the cultivated natural capital of a plantation forests, however, remains useful, not only for the lumber but as a way of easing the pressure of lumber interests on the remaining true natural capital of nonplantation forests.

Marketed natural capital can, subject to the important social corrections for common property and myopic discounting, be left to the market. Nonmarketed natural capital, both renewable and nonrenewable, presents the most troublesome category for future economic planning. Remaining natural forests should in many cases be treated as nonmarketed natural capital, and only replanted areas treated as marketed natural capital. In neoclassical terms the external benefits of remaining natural forests might be considered "infinite," thus removing them from market competition with other (inferior) uses. Most neoclassical economists, however, have a strong aversion to any imputation of an "infinite" or prohibitive price to anything.

### Policies for a Full World

In this new full-world era investment must shift from manmade capital accumulation toward natural capital preservation and restoration. Also technology should be aimed at increasing the productivity of natural capital rather than that of manmade capital. If these two things do not happen then we will be behaving *uneconomically*, in the most orthodox sense of the word. Market forces could bring about this change if the price of natural capital were to rise as it became more scarce. What keeps the price from rising? In most cases natural capital is unowned and consequently nonmarketed. Therefore it has no explicit price and is exploited as if its price were zero. Even where prices exist for natural capital, the market tends to be myopic and excessively discounts the costs of future scarcity, especially when under the influence of economists who teach that accumulating capital is a near-perfect substitute for depleting natural resources.

Natural capital productivity grows by means of three things: increasing the flow (net growth) of natural resources per unit of natural stock (which is limited by biological growth rates); increasing product output per unit of resource input (limited by mass balance); and, especially, increasing the end-use efficiency with which the resulting product yields services to the final user (limited by technology). I have already argued that complementarity severely limits what we should expect from the second, and complex ecological interrelations and the law of conservation of matter and energy will limit the increase from the first of these. Therefore I shall focus on the third option.

The above factors limit productivity from the supply side. From the demand side tastes may provide a more stringent limit to the economic productivity of natural capital than that of biological productivity. Game ranching and fruit and nut gathering in a natural tropical forest may, in terms of biomass, be more productive than cattle ranching. But undeveloped tastes for game meat and tropical fruit may make this use less profitable than the biologically less productive use. In this case a change in tastes would increase the biological productivity with which the land is used.

Because capitalists own manmade capital, we can expect that it will be maintained with an interest to increasing its productivity. Labor power, which is a stock that yields the useful services of labor, can be treated in the same way as manmade capital. Labor power is manmade and owned by the laborer who has an interest in maintaining it and enhancing its productivity. But nonmarketed natural capital (the water cycle, the ozone layer, the atmosphere) is not subject to ownership, and no self-interested social class can be relied upon to protect it from overexploitation.

If development economists and multilateral development banks were to accept my thesis, what policy implications would follow? The role of multilateral development banks in the new era would be increasingly to make investments that replenished the stock and increased the productivity of natural capital. In the past development investments have largely aimed at increasing the stock and productivity of manmade capital. Instead of investing mainly in sawmills, fishing boats, and refineries, therefore, development banks would now invest in reforestation, restocking of fish populations, and renewable substitutes for dwindling reserves of petroleum. The latter should include investment in energy efficiency, for it is impossible to restock petroleum deposits. Natural capacity to absorb wastes also constitutes a vital resource, so investments that preserve that capacity, like pollution reduction, also increase in priority. For marketed natural capital such policies will not represent a revolutionary change. Even for nonmarketed natural capital, there are economic development agencies with experience in investing in complementary public goods such as education, legal systems, public infrastructure, and population control. Investments in limiting the rate of growth of the human population are essential to managing a world that has become relatively full. Like manmade capital, manmade labor power is also complementary with natural resources, and its growth can increase demand for natural resources beyond the capacity of natural capital to supply sustainably.

Perhaps the clearest policy implication of the full-world thesis is that the level of per-capita resource use of the rich countries cannot be generalized to the poor, given the current world population. Present total resource-use levels are already unsustainable, and multiplying them by a factor of from five to ten, as envisaged in the Brundtland report (World Commission on Environment and Development 1987), is ecologically impossible. As a policy of growth becomes less possible the importance of redistribution and population control as measures to combat poverty increase correspondingly. In a full world both human numbers and per-capita resource use must be constrained. Poor countries cannot cut per-capita resource use—indeed they must increase it to reach a sufficiency—so their focus must center on population control. Rich countries can cut both, and those that have already reached demographic equilibrium need to focus on limiting per-capita consumption, in order to make resources available to transfer to poorer countries, to bring them up to sufficiency. Investment in the areas of population control and redistribution therefore increases in priority for development agencies.

Investing in natural (nonmarketed) capital remains essentially an infrastructure investment on a grand scale and in the most fundamental sense of infrastructure—to wit, the biophysical infrastructure of the entire human niche, not just the within-niche public investments that support the productivity of private investments. We now need investments in biophysical infrastructure ("infra-infrastructure") to maintain the productivity of all previous economic investments in manmade capital, public or private, by rebuilding the remaining natural capital stocks that have come to be limiting. Indeed, in the new era the World Bank's official name, the International Bank for Reconstruction and Development, should emphasize the word *reconstruction* and redefine it to refer to reconstruction of natural capital devastated by rapacious "development," rather than reconstruction of manmade capital in a Europe devastated by World War II, its historical meaning. Because our ability actually to recreate natural capital remains limited, such investments will have to be indirect. We must, for example, conserve the remaining natural capital and encourage its natural growth by reducing our level of current exploitation. This includes investing in projects that relieve the pressure on these natural capital stocks by expanding cultivated natural capital (like using plantation forests to relieve pressure on natural forests) and increasing end-use efficiency of products.

The difficulty with infrastructure investments is that their productivity shows up in the enhanced return to other investments and is therefore difficult both to calculate and to collect for loan repayment. Also, in the present context these ecological infrastructure investments are defensive and restorative in nature—that is, they will protect existing rates of return from falling more rapidly than they would otherwise, rather than raise their rate of return to a higher level. Although this circumstance will dampen political enthusiasm for such investments, it does not alter the economic logic favoring them. Past high rates of return to manmade capital were possible only with unsustainable rates of use of natural resources and consequent (uncounted) liquidation of natural capital.

We are now learning to deduct natural capital liquidation from our measure of national income (See Ahmad et al. 1989). The new era of sustainable development will not permit natural capital liquidation to count as income and will consequently require that we become accustomed to lower rates of return on manmade capital—rates on the order of magnitude of the biological growth rates of natural capital, for they will become the limiting factor. Once investments in natural capital have resulted in equilibrium stocks that are maintained but not expanded (yielding a constant total resource flow), then all further increase in economic welfare would have to come from increases in pure efficiency. Efficiency increases result from improvements in technology and clarification of priorities. Certainly investments are being made in improving biological growth rates, and the advent of genetic engineering will add greatly to this thrust. Experience to date with such movements as the green revolution, however, indicates that higher biological yield rates usually require the sacrifice of some other useful quality, like disease resistance, flavor, or strength of stalk. In any case, the law of conservation of matter and energy cannot be evaded by genetics. More food from a plant or animal implies either more inputs or less matter-energy going to the nonfood structures and functions of the organism. To avoid ecological backlashes will require leadership and clarity of purpose on the part of development agencies. To carry the arguments for infrastructure investments into the area of biophysical infrastructure or natural capital replenishment will require new thinking by development economists. Because much natural capital is not only public but globally public in nature, the United Nations seems best suited to take a leadership role.

Consider some specific cases of biospheric infrastructure investments and the difficulties they present. First, a largely deforested country will need reforestation to keep the complementary manmade capital of sawmills (carpentry, cabinetry skills, and others) from losing their value. Of course, the deforested country could for a time resort to importing logs. But that would provide a temporary relief at best. To protect the manmade capital of dams from the silting up the lakes behind them, the water catchment areas feeding the lakes must be reforested or original forests protected to prevent erosion and siltation. Agricultural investments depending on irrigation can become worthless without forested water catchment areas that recharge aquifers.

Second, on a global level enormous stocks of manmade capital and natural capital are threatened by depletion of the ozone layer, although the exact consequences are too uncertain to be predicted. The greenhouse effect is a threat to the value of all coastally located and climatically dependent capital, be it manmade (port cities, wharves, beach resorts) or natural (estuarine breeding grounds for fish and shrimp). And if the natural capital of fish populations diminishes because of the loss of breeding grounds, then the value of the manmade capital of fishing boats and canneries will also decrease, as will the specialized human capital devoted to fishing, canning, and related industries. We have begun to adjust national accounts for the liquidation of natural capital but have not yet recognized that the value of complementary manmade capital must also be devalued, as the natural capital that it was designed to exploit disappears. Eventually the market will automatically lower the valuation of fishing boats as fish disappear, so perhaps no accounting adjustments are called for. But *ex ante* policy adjustments aimed at avoiding the *ex post* devaluation of complementary manmade capital, whether by market or accountant, are certainly called for.

Although there is as yet no indication of the degree to which development economists agree with my thesis, three U.N. agencies (World Bank, UNEP, and UNDP) have nevertheless embarked on a project, however exploratory and modest, of biospheric infrastructure investment, known as the Global Environment Facility. The facility would provide concessional funding for programs investing in the preservation or enhancement of four classes of biospheric infrastructure or nonmarketed natural capital. These are protection of the ozone layer; reduction of greenhouse gas emissions; protection of international water resources; and protection of biodiversity. If I am right in my argument, then investments of this type should eventually become important in the lending portfolios of development banks. My thesis does at least provide theoretical justification and guidance for present efforts to shape the Global Environment Facility and its likely extensions. It would seem that the "new era" thesis merits serious discussion, both inside and outside the multilateral development banks, for our practical policy response to the reality of the new era has already outrun our theoretical understanding of it.

**References**

The views presented here are those of the author and should in no way be attributed to the World Bank. I am grateful to S. El Serafy, P. Ehrlich, R. Goodland, B. Hannon, G. Lozada, R. Overby, S. Postel, B. von Droste, and P. Dogse for helpful comments.

Ahmad, Y., S. El Serafy, and E. Lutz, eds. 1989. *Environmental Accounting for Sustainable Development*. Washington, D.C.: World Bank.

Boulding, K. 1964. *The Meaning of the Twentieth Century*. New York: Harper and Row.

Daly, H. E., and J. B. Cobb. 1989. *For the Common Good*. Boston: Beacon Press.

Georgescu-Roegen, N. 1971. *The Entropy Law and the Economic Process*. Cambridge, Mass.: Harvard University Press.

Goodland, R. 1991. "The Case That the World Has Reached Limits." *Environmentally Sustainable Economic Development: Building on Brundtland*, ed.

R. Goodland, H. Daly, S. El Serafy, and B. von Droste, 15–27. Washington, D.C.: World Bank.

Takayama, A. 1985. *Mathematical Economics*. New York: Cambridge University Press.

Vitousek, P. 1986. "Human Appropriation of the Products of Photosynthesis." *Bioscience* 34(6):368–373.

World Commission on Environment and Development. 1987. *Our Common Future* (the Brundtland report). Oxford: Oxford University Press.

# 5
## Government Policy, Economics, and the Forest Sector

**Robert Repetto**

**G**overnments, many of which are committed in principle to conservation and wise resource use, are aggravating the loss of the forests under their stewardship through mistaken policies. Such policies, by and large, were adopted for worthy objectives: industrial or agricultural growth, regional development, job creation, or poverty alleviation. But such objectives typically have not been realized or have been attained only at excessive cost.

Throughout the world, governments largely determine how forests should be used. In the industrialized countries, a substantial percentage of remaining forests are on public lands. In the United States, for example, the National Forest System includes $77 \times 10^6$ ha, more than a quarter of all forest land. National forests now provide approximately one third of the soft wood harvest in the United States. And according to a comprehensive FAO assessment, in the Third World over 80 percent of the closed forest area is public land (Lanly 1982). Governments have taken over the authority and responsibility for managing forests from indigenous communities, which traditionally used them in accordance with their own laws.

Government policies influence even the use of private forests, intentionally or not. Because commercial forestry involves holding a growing asset for long periods, returns to private investors are sensitive to credit costs, inflation, taxes on land and capital assets, and other economic factors greatly affected by government policy. And because some forest lands can be used for agricultural or other purposes, government policies that stimulate expansion of these competing land uses threaten forest area.

Although the area of forest cover in industrialized countries has not changed significantly in recent years, that in developing countries was shrinking in the late 1970s by about $11 \times 10^6$ hectares each year. This chapter therefore focuses on forests in the Third World, although some of the dynamics among forests, government policy, and the broader economy hold for industrial countries as well.

### Forests and Agriculture in the Broader Economy

In developing regions of Asia, Africa, and Latin America, forests and woodlands cover 20 to 25 percent of the total land area. Thus, forces affecting the demand for agricultural, industrial, or residential land will probably affect forested area. Agricultural land, including cropland and pasture, accounts for an additional 20 percent of total land area, so the land that has been deforested yet not used for agricultural purposes is particularly important. Most of the deforested land by far has gone into agricultural production.

In broad terms, changes in demand for agricultural land depend on the growth of demand for agricultural output and on shifts in the mix of inputs used in agricultural production. In turn, changes in domestic agricultural demand depend on growth in population, per-capita income, and distribution of that income. All three, of course, are strongly interrelated and depend on macroeconomic policies. Countries in which population and per-capita income grow quickly and with a fairly even distribution usually experience rapidly rising domestic agricultural demands. With improving per-capita income, demand increases more slowly for cereals than for animal products and other foods, with further implications for agricultural land use.

On the supply side, the proportions with which land, labor, and capital are used in agricultural production basically depend on their relative scarcities. Sparsely populated, low-income countries have typically adopted extensive agricultural production systems, including free grazing of livestock, shifting cultivation of foodcrops, and production of plantation crops (Binswanger and Pingali 1984). Because most of the remaining forests are in sparsely populated regions, these closely competing agricultural production systems account for most of the forest conversion.

As long as the cost of converting adequate reserves of grasslands and forests to agricultural production remains roughly constant, even as demands and rural labor supplies increase, little change in production systems or in the amount of land being converted occurs (Boserup 1981). Agricultural expansion occurs through an increase of acreage planted, as was the case throughout the nineteenth century and up to World War II in most developing countries, and is still the case in sparsely populated regions.

Exploiting the supply of land at the frontier of cultivation usually involves privatizing forests and grasslands held under communal tenures or nominally controlled by national governments. Because those governments rarely enforce the traditional land rights of forest-dwelling communities and typically encourage or acquiesce in land acquisition by outside settlers and entrepreneurs, land at the frontier is largely an open-access resource. Neither traditional users nor government agencies can effectively exclude immigrants. Because of this market failure, the opportunity costs of land conversion are inadequately reflected in the supply price of forested land. In other words, the supply price of forested land does not adequately reflect the value of services performed by forests while they are intact. Land rights and tenurial institutions strongly influence the rate of deforestation.

Even with the rising costs of land conversion, macroeconomic conditions greatly influence changes in factor scarcities in agriculture. Land scarcities at the extensive margin are affected by the overall distribution of agricultural landholdings. In many countries, highly concentrated landownership leads to less intensive use of the most productive agricultural land and a greater demand for land on the agricultural frontier. Favorable tax treatment of agricultural land or income and agricultural credit and input subsidies can lead to the consolidation of landholdings.

Rural labor supply depends not only on underlying population growth rates but also on the rate of growth of urban labor demand, which depends in turn on the rate and pattern of industrialization (Ranis and Fei 1961). Movements of agricultural labor to the frontier are affected by changes in urban labor absorption and by capital-labor substitutions within the agricultural sector.

Capital availability in agriculture relies on the distribution of income and terms of trade between agriculture and industry and also on capital and product market distortions that artificially alter relative rates of return to investments in the two sectors (Mellor 1976). Also, complementary public investments in rural infrastructure and agricultural research influence the level of private agricultural investment.

The effects of macroeconomic policy on the agricultural land-use frontier are thus strong and varied. But on the forested side of the frontier, policy influences are also at work. Domestic demand for forest products increases with population and income growth. Wood for fuel, pulp, construction, and industrial uses is the principal forest resource, although a wide variety of nonwood products are also harvested. The income elasticity of demand for pulp and industrial wood is high in rapidly urbanizing developing countries, but fuelwood, including charcoal, still accounts for the greatest volume of wood extracted from open forests in semiarid areas and forest fringes. Fuelwood demand depends also on the relative cost of substitute fuels, such as kerosene, electricity, and other biomass residues.

The market failure arising from insecure tenurial rights dominates the supply of most forest products in developing countries. Where current demands for wood exceed annual forest growth, they are met by cutting into the stock at rates that implicitly heavily discount future values. Insecure tenures discourage conservation of forest stocks for future use, whether the insecurity arises from the breakdown of communal resource management traditions (Jodha 1990), lack of enforcement of government restrictions on the use of public forests, or irregularities in the administration of timber harvesting rights (Repetto and Gillis 1988). Even in countries where natural forests are severely depleted, private investment in reforestation is depressed, because market prices rarely reflect the user costs of wood harvested from openaccess natural forests.

Industrial timber is typically harvested either under long-term concessions from the national government (in Asia and Africa) or from privately owned forest lands (in Central and South America). Because timber is a long-lived asset, industrial wood supply is greatly influenced not only by insecurities of tenure but also by capital market conditions. Most industrial timber in developing countries still comes from old-growth natural forests, although in an increasing number of countries these stocks are rapidly being exhausted. The timber owner's opportunity cost of capital influences his or her decisions with regard to delaying the initial harvest, leaving adequate stocks for regeneration, managing the forest for a subsequent rotation, and allowing sufficient time for regeneration before reentering a stand (Chang 1981).

In few developing countries are national capital markets open to or reasonably integrated with world capital markets. Fluctuating real exchange rates, trade and investment restrictions, and legal and institutional barriers restrict international capital flows. Moreover, macroeconomic and sectoral policies strongly influence the relative rates of return to investments in various sectors, usually inflating industrial profits at the expense of the agricultural, forest, fishing, and mining sectors. Consequently, large-scale holders of timber rights with profitable investment opportunities in industry or elsewhere may have high opportunity costs of capital and hence strong incentives to mine the forest for short-term profits.

Other timber supply conditions reinforce these incentives. First, forest proprietors cannot profit economically from the ecological benefits old-growth forests produce in the form of wildlife habitat, hydrological cycling, and climate regulation. This market failure biases timber management decisions toward shorter rotations, smaller residual stands, and lower levels of investment in forest management (Hartman 1976).

Second, in countries where logging by private firms on public lands is carried out under concession agreements and licenses, government failure to capture the resource rent from old-growth timber effectively makes the forest a common resource. Royalties and taxes collect only a fraction of the stumpage value, so logging firms engage in rent-seeking behavior, accumulating large concession areas, lest they be awarded to others, and then selectively harvesting them for the most valuable logs. Restrictions against stockpiling concessions and short concession periods reinforce loggers' incentives to maximize short-term rents. In countries where old-growth forests are transferred from the public domain to private ownership, laws that award land titles on evidence that forest land has been improved, usually by clearance and conversion to other uses, have a similar effect on forest management decisions.

So far in this discussion, although I have suggested many macroeconomic policy linkages, I have still neglected the role of international trade. Countries at early stages of development export mostly primary commodities, with little processing. Logs have been (and still are) important exports from many forest-rich countries. With development, the composition of exports changes to include more added value in the form of processed commodities and labor-intensive manufactures. Over time, the capital and skill content of exports from successful developing countries increases. The experience of such successful, outward-looking economies as Korea and Thailand follow this pattern.

Outward-looking trade policies can create strong export demands for resource-based commodities, including forest and competing land-intensive products like plantation crops and animal products. Unless export or other taxes capture the rents from these infra-marginal exports, thus limiting the potential of suppliers to exploit these resources, the amount supplied can expand rapidly, leading to depletion of forest stocks and conversion of forest land to agricultural uses. Protectionist policies designed to promote local timber processing and forest-based industrialization can build a large, technically inefficient milling industry with sizable demands on raw materials. Market failures surrounding forest conversion, combined with strong external demand, promote rapid deforestation (Repetto and Holmes 1983). In time, in outward-looking economies, the shifting comparative advantage toward manufactures and rising supply costs of land-intensive exports tend to reduce the pressure of external demand on remaining forest stocks.

In the short run, inward-looking trade regimes, which heavily protect domestic import-substituting industries and discourage exports, may reduce external pressures on forests. Overvalued exchange rates and negative effective rates of protection reduce the rents from primary product exports. Over the longer run, however, inward-looking trade regimes may actually result in greater pressures on forest resources. Adverse incentives prevent the development of manufactured exports, so that countries following such policies remain heavily dependent for foreign exchange earnings on resource-based commodities having strong comparative advantage and tend to exploit them heavily. Heavy industrial protection, by inflating industrial profit rates in the short run, raises the opportunity cost of capital tied up in forest stocks. Moreover, while the slower aggregate growth rates associated with inward-looking policies limit demands for agricultural and forest products, the lower growth rates of the industrial sector also limit the transfer of labor from agriculture into urban occupations. Rural underemployment accelerates migration to agricultural frontiers.

Macroeconomic policies have pervasive influences on the use and conservation of forest resources. One way of visualizing these influences is as a set of concentric circles centered on the forest. At the hub are policies directly affecting timber and forest management, such as forest revenue structures, tenurial institutions governing privatization of forest land and enforcement of traditional use rights, and administration of timber harvesting concessions.

The next circle contains policies directly influencing the demand for forest products, such as trade and investment incentives to promote wood-using industries, and energy pricing policies toward fuelwood substitutes. In a third circle might fall agricultural policies directly affecting movements of the agricultural frontier and the rate of conversion of forested land. These include credit, tax, and pricing incentives for land-intensive plantations and ranches. They include all policies that lead to supply expansion at the extensive rather than the intensive margin, such as those that increase the concentration of landholdings, lower the marginal returns to labor and capital in agriculture, or direct public infrastructure spending toward frontier expansion.

In a fourth, broad outer circle lie macroeconomic policies that at first seem unrelated to forest management but actually have powerful impacts. In this circle are policies that retard the demographic transition by perpetuating poverty and discrimination against women. Also here are trade and investment policies that decrease the flow of migration from rural to urban areas by preventing the rapid growth of labor-intensive industries and ancillary employment opportunities. No less important are policies that shorten investors' time horizons by distorting capital markets or by encouraging rent-seeking behavior.

These policy influences can now be discussed only qualitatively and piecemeal. There are no macroeconomic models or analytical frameworks that effectively link forest exploitation and forest land conversion to this array of sectoral and macroeconomic policies. A start has been made in modifying conventional macroeconomic accounting frameworks to incorporate changes in natural resource stocks in an economically satisfactory way (Repetto et al. 1989). Only a few countries in the developing world, however, are empirically estimating natural resource accounts so as to be compatible with the national income accounting systems. An appropriate set of accounts would provide a useful foundation on which analytical frameworks linking macroeconomic policy and deforestation could be constructed.

We have little empirical research that tests the linkages sketched out in the preceding pages. It would be overly ambitious at this stage to attempt an overall assessment of the whole range of macroeconomic policy influences simultaneously. Rather, well-focused research studies illuminating subsets of these policy linkages would be valuable building blocks. The preceding pages, in effect, sketch out an extensive research agenda. Now that national and international development agencies are coming to grips with the problems of deforestation and trying to formulate policy changes to control it, the results of such studies would be timely and useful. The research must, of course, specify individual countries, not only because national policymakers are more interested in information about their own options but also because despite broad similarities policy linkages differ from country to country. For this reason an international network of collaborative research could stimulate and implement research on this agenda.

### Economic Losses Resulting from Deforestation

Tropical forests are now being destroyed much more extensively than they were a decade ago, unless rates of deforestation for that period were greatly underestimated. Compared to the most careful assessments of forest disturbance and clearance available in the early 1980s, more recent estimates for a number of countries, based on satellite imaging and ground surveys, show significantly higher rates of deforestation. In India, for example, studies by the National Remote Sensing Agency resulted in an increase in the estimated deforestation rate during the early 1980s to $1.3 \times 10^6$ ha per year, nine times the earlier FAO estimate of $0.147 \times 10^6$ ha. It was found that large areas of land legally designated as forest were virtually treeless. In Brazil, the FAO estimate of $2.5 \times 10^6$ ha annually for the early 1980s, which was based on partial Landsat surveys and other sources, increased to $8 \times 10^6$ ha per year for 1985–1988, based on more recent interpretations of satellite data. (Setzer et al. 1988; Malingreau and Tucker 1988). In 1987, $20 \times 10^6$ ha in the Amazon (much of it grassland and pasture) were in flames. The 1989 estimate for deforestation in Brazil was $4.6 \times 10^6$ ha.

Deforestation at this rate poses extreme risks to natural systems. The consequent release of carbon to the atmosphere probably contributes 15 to 30 percent of annual global carbon emissions, a substantial contribution to the buildup of greenhouse gases (Woodwell et al. 1983; Detwiler and Hall 1988; Houghton et al. 1987) Moreover, loss of tropical forests is rapidly eliminating the habitat of large numbers of plant and animal species. Up to half of the world's plant and animal species inhabit tropical forests, and in ten biotically rich and severely threatened regions totalling just 3.5 percent of the remaining tropical forest area, 7 percent of all plant species will probably go extinct by the end of the century if current trends continue (Myers 1988).

The dismay of scientists and environmentalists over this destruction is now shared by leaders of the world's governments. The Paris summit meeting of heads of governments in the Group of Seven countries (Canada, France, Germany, Italy, Japan, United Kingdom, and United States) in July 1989 declared that "preserving the tropical forests is an urgent need for the world as a whole. . . . We express our readiness to assist the efforts of nations with tropical forests through financial and technical cooperation, and in international organizations." Deforestation has reached agendas of summit meetings on the global economy as well.

A sense of crisis also emerges in the tropics, as once ample forest resources disappear. An asset capable of yielding rich returns indefinitely has been badly depleted. In Thailand, the government banned commercial logging, over the protests of influential concession-holders, when surveys showed that between 1985 and 1988 forest cover had declined from 29 to 19 percent of the land area, and landslides from deforested hillsides cost forty thousand people their homes (*Asiaweek,* 10 March 1989). In the Philippines, undisturbed dipterocarp forests have shrunk from $16 \times 10^6$ ha in 1960 to less than a million hectares still standing in remote hill regions (Boado 1988). Logging has been suspended in most provinces, and mills in the Philippines are either closing or importing logs from Sabah and Sarawak. Mills in the once-rich Indonesian production centers of Sumatra and Kalimantan are also experiencing shortages of accessible high-quality timber and importing logs from Sabah, Sarawak, and Irian Jaya. Indonesia's ambitious timber development plans are now faced with potential resource shortages. Sabah and Sarawak, currently the major sources of logs in Asia, are harvesting almost twice the sustained yield of their forests and will also be rapidly depleted (Burgess 1988). In West Africa, Central America, and China as well, the loss of most mature forests has depressed incomes, foreign exchange earnings, and employment from forest-based industries.

Although the issues are complicated by global environmental concerns voiced primarily in developed countries, governments of most tropical countries are coming to realize that to them rapid deforestation represents a severe economic loss, a waste of valuable resources. In the Ivory Coast, for example, where forest cover has decreased by 75 percent since 1960, an estimated $200 \times 10^6$ m³ of commercial timber has simply been burned to clear the land, a loss of perhaps $5 billion. The Forest Department of Ghana believes that in that country, where 80 percent of the forests have disappeared, only 15 percent of them were harvested before the land was cleared (Rietbergen 1988). And in Brazil, where little of the timber is extracted before a forest is burned, the resulting loss in commercial timber reaches approximately $2.5 billion *annually* (Schmidt 1989). That is a quarter of Brazil's annual net debt-servicing payment on its external debt.

Burning valuable timber while clearing forests for other land-use purposes represents only one obvious kind of economic wastage. Other losses spring from the extremely short timeframes in which tropical forests are exploited. Loggers destroy enormous quantities of valuable timber in tropical forests through careless use of equipment, failure to cut away vines before felling trees, and other practices. If loggers extract 10 percent of the timber, by selecting mature trees of the most valuable species, they typically destroy at least half the remaining stock, including immature trees of the same valuable species, as well as harvestable stock of less well-known varieties. They thus reduce harvesting costs, but the residual stand and subsequent harvests are impoverished.

Loggers often reenter logged areas to extract more timber before stands have recovered, inflicting heavy damage on residual trees each time and making regeneration impossible. In Ghana and the Ivory Coast, stands have been reentered as often as three times in 10–15 years, when concessionaires obtained sales contracts for logs of lesser-known species (Rietbergen 1988). Dipterocarp forests in the Philippines have been exploited on a clearly unsustainable cutting cycle of 6–8 years. This also reflects an extremely high time discount rate by concessionaires with a high opportunity cost of capital or little confidence that they will enjoy the benefits of future harvests, or both.

A recent study commissioned by the International Tropical Timber Organization (ITTO) found that not even 0.1 percent of remaining tropical forests are being actively managed for sustained productivity (Poore 1988). Destructive logging practices on short cutting cycles result in severely depleted timber stands. Moreover, production forests in most countries are left virtually unprotected after the harvest from encroachment by shifting cultivators and are thus exposed to burning and clearing (Wyatt-Smith, in Mergen and Vincent 1987). This is additional evidence of loggers' high discount rates. Surveys in the Amazon demonstrate that deforestation proceeds rapidly where roads for logging or other purposes have opened up a region, but remains minimal elsewhere.

The biological degradation of tropical forests carries an increasingly high economic price tag. The timber cost alone has been unexpectedly large, for tropical timber prices have bucked the general downward trend of declining commodity prices, and many previously uncommercial species now find ready markets. In West Africa, for example, the price of previously neglected logs is now on a level with prime varieties. Countries that had earlier extracted as few as two or three trees per hectare deforested, destroying the rest as uncommercial, now regret their shortsightedness. The upward trend in tropical timber prices will likely continue, as supplies are depleted over the next decade in Asia, Central America, and West Africa. Consequently, the timber in the Amazon basin, which is now being recklessly burned, will become increasingly valuable. For old-growth timber, an asset with a low rate of biological appreciation, expected price increases are necessary if stocks are to be held for future use.

But potential timber values are by no means the only economic losses that deforested countries suffer. Probably 70 percent of the wood harvested in tropical countries is used locally, and as forests recede fuelwood shortages increase. Other forest products become unavailable to local residents, including bushmeat, fruits, oils, nuts, sweeteners, resins, tannins, fibers, construction materials, a wide range of medicinal compounds, and such salable products as skins, feathers, and live animals. In Indonesia, the value of only those nontimber forest products that reached the export market came to $123 million by 1986 (HIID 1988). Many such nontimber products are exploited as open-access resources. Nonetheless, recent studies have shown that the capitalized value of the income derived from nontimber forest products, which can be extracted sustainably, may greatly exceed that of the timber harvest (Peters et al. 1989). Moreover, the incomes so derived support local residents, while the profits from timber exploitation are typically captured by distant elite power groups or foreign corporations. Timber operations have therefore sparked violent protests by indigenous communities in Sarawak, the Philippines, and other countries. Were nontimber values adequately reflected in timber management decisions, we can safely assume that initial harvests would be delayed, residual stands protected, and reentry deferred to a greater extent than is now common.

Deforestation often has severe environmental impact on soils, water quality, and even local climate. Shallow, easily leached soils are damaged by heavy equipment, and, when exposed to heavy tropical rains, they can quickly erode or lose remaining nutrients. Studies in Ghana showed that eliminating savanna forest raised soil erosion rates from less than 1 to more than 100 tons per hectare, with a nutrient loss 40 percent higher than comes from the average annual chemical fertilizer application (World Bank 1988). Plentiful riverine fisheries have been damaged by increased sedimentation from deforestation in floodplains which provide critical seasonal habitat. Large-scale tropical deforestation interrupts moisture recycling, reducing rainfall and raising soil temperatures, and perhaps leading to long-term ecological changes (Salati and Vose 1983). These environmental values are currently external to loggers' decisions, but they make forest conservation a rational economic strategy.

Moreover, deforestation has often accompanied shifts to economically and environmentally inferior land uses, such as cattle ranching and inappropriate modes of agriculture. Given the quick loss of productivity and low carrying capacity of pastures in the rapidly deforesting Brazilian State of Acre, for example, the net present value of revenues per hectare from collecting wild rubber and brazil nuts was found to be four times that of cattle ranching (Prickett 1988). In Guatemala, studies have shown that sustained forest management for nontimber and timber production is economically superior to slash-and-burn agriculture. Tax and credit incentives, however, as tenurial rules requiring alteration of the natural forest in order to obtain or confirm private land titles, and land speculation in inflation-prone countries promote such inferior land uses.

### Policy Influences on Tropical Forests

Above all, governments of developing countries, which are the proprietors of at least 80 percent of the closed tropical forests, have not put an adequate value on the resource. As proprietors, they could capture the entire resource value except for the cost of labor and capital employed in managing and harvesting by charging sufficient royalties and taxes or by selling harvesting rights to the highest bidders. Instead, with few exceptions, governments have allowed most of these resource rents to flow to timber concessionaires and speculators, often linked to foreign enterprises. Governments have created these windfalls by keeping royalties and fees charged to timber concession-holders low, reducing export taxes on processed timber to stimulate domestic industry, and even granting income tax holidays to logging companies.

Governments have failed even to enforce the official charges effectively. Consequently, few governments of tropical countries for which data exist have succeeded either in limiting timber exploiters to a normal rate of profit or in capturing the value of the forest resource for the public treasury. (The same is true, incidentally, of the governments of many temperate countries, including the United States, Canada, and Australia.)

This has had most unfortunate consequences. It has sparked timber booms throughout the tropics, drawing both domestic and foreign entrepreneurs—many with little forestry experience—into the search for quick fortunes. Under this pressure, governments have awarded timber concessions covering areas far greater than they could effectively supervise or manage, sometimes overlapping protected areas and national parks.

While sacrificing enormous sums in potential forest revenues, governments in the tropics are failing to invest enough in stewardship and management of their resource. Although forestry codes and stipulations in concession agreements for many countries seem adequate to ensure sustained productivity over at least several growing cycles, given the evidence of experimental research (Hadley 1988), natural forests are almost nowhere being managed to achieve that goal (Rietbergen 1988; Burgess 1988; Mergen and Vincent 1987; Poore 1988).

Ineffective government supervision is compounded by the perverse incentives created for timber companies by the terms of concession agreements, which discourage any possible interest companies might have in sustained yield management. Most agreements run for twenty years or less, and many for five years or less, although ecologically sound management prescribes twenty-five- to thirty-five-year intervals between successive harvests in selective cutting systems and longer intervals in monocyclic systems. Concession-holders therefore have little reason to care whether productivity is maintained for future harvests. Even though longer concession periods are undoubtedly insufficient to lower loggers' time discount rates, they may be necessary.

Moreover, forest revenue systems based on relatively undifferentiated fees levied on the volume of wood extracted encourage loggers to take only the highest value logs at minimum cost, which leads to "highgrading" the timber (harvesting the best trees while disregarding inferior trees) over large areas and extensive damage to residual stands.

Trees with stumpage value less than the royalty rate appear worthless to the concession-holder, who destroys them with impunity. Royalties based on the size of the concession and the total merchantable timber it contains encourage more complete use of the timber resource within a smaller harvesting area. Ad valorem royalties also encourage fuller utilization.

Distorted incentives affect the efficiency of wood-processing industries as well. Log-producing countries have had to provide strong incentives to local mills to overcome high rates of protection against processed wood imports into Japan and Europe. Extreme measures, such as log export bans and log export quotas based on volumes processed domestically have created inefficient local industries, sometimes set up only to preserve valuable log export rights.

Countries sheltering inefficient processing industries, although they are trying to increase local added value and employment, can incur heavy economic and fiscal losses. In the Philippines, for example, each log exported as plywood is worth $100–110 per cubic meter less than it would be if it were exported without processing or as sawn timber, while the government sacrifices more than $20 million annually in export taxes to encourage plywood exports. In 1987, the Philippines dissipated more than two-thirds of their potential resource rents from the tropical timber harvest in inefficient processing induced by excessive protection (Repetto 1988).

Industrial countries have contributed to, and profited from, problems of forest policy in the tropics. Japan is the largest importer of tropical timber, accounting for 29 percent of world trade in 1986, roughly the same share as that of the EEC. Japanese imports, mostly of unprocessed logs (unlike those of the EEC) were 30 percent higher in 1987, largely because of the construction boom. Most tropical hardwood imports are used in Japan for construction plywood, primarily as disposable forms to mold concrete, for which tropical woods were adopted because of their cheapness (Nectoux and Kuroda 1989).

European and U.S. companies have long held interests in logging and processing enterprises, especially in tropical Africa and Latin America, but Japanese business now heavily outweighs its rivals in the tropical timber trade. The large Japanese trading companies are involved in all stages of exploitation: as partners and financiers of logging concessionaires, as exporters and importers, and as processors and distributors. Japanese firms have shifted their attention from the Philippines to Indonesia to Sabah and Sarawak as log supplies are successively depleted, and they are now interested in Amazonian forests as a potential future source of raw material (Nectoux and Kuroda 1989). They have shown little interest in sustained management of their holdings. Instead, highly leveraged operations harvest as much and as quickly as possible to pay their financing charges (Nectoux and Kuroda 1989).

Of course, policies of industrial countries influence economic development in tropical countries. Much of the Third World has suffered from economic stagnation and decline as a consequence of recession in the industrialized countries in the early 1980s, high real interest rates, sharp reductions in net capital flows to developing countries, and trade protection in the industrial countries. The connection must be pointed out between economic stagnation leading to deforestation in the Third World and policies in industrial countries that restrict capital flow to developing countries and markets for developing country exports.

More specifically, trade barriers in the forest-products sector erected by industrial countries have been partially responsible for inappropriate investments and patterns of exploitation in developing nations' forest industries. Whether within the context of the General Agreement on Tariffs and Trade (GATT), the International Tropical Timber Agreement (ITTA), or some other international forum, negotiations between exporting and importing countries need to reduce tariff escalation and non-tariff barriers to processed-wood imports from tropical countries and find some way to rationalize incentives to forest industries in the Third World.

Both experience and analysis reinforce the argument that deforestation has not been a path to development but is in most tropical countries a costly drain of increasingly valuable resources. Recognition is also growing that deforestation need not be inevitable in developing countries, being rather the consequence of poor stewardship, inappropriate local, national, and international policies, and inattention to problems outside the forest sector.

Fortunately, there are many indications of a new approach to policies for tropical forests that reflects the increasing awareness of their national and global significance. Many governments of developing countries are taking steps to capture the resource rents that have motivated the despoliation of tropical forests. A number of countries are now strengthening their forest management capabilities, with the help of development assistance agencies. Yet these agencies are still financing projects destructive to tropical forests. Such projects should be supplanted by others designed to improve forest management and greatly expand the pace of reforestation efforts.

A great deal remains that the world outside the tropics can do. Consumption patterns, such using tropical hardwoods for disposable concrete molds, and some business practices of industrial countries contribute to deforestation. Change in these areas would do much to help tropical countries control deforestation and stem economic losses. Trade policies in industrial countries, too, need to be reformulated in light of their effect on tropical deforestation.

The scope for international cooperation to halt the destruction of tropical forests is large. The World Bank's Global Environmental Facility, the principles on forest conservation and use currently being formulated in the United Nations Conference on Environment and Development process, and the international convention on global climate change adopted in May 1992 could be powerful mechanisms for international cooperation. They are also the mechanisms through which northern countries with interests in the preservation of the global environment can share in the costs of actions taken by developing countries. New forms of international cooperation, such as an international convention on forests, would reflect our growing awareness that disappearing forests, including tropical forests, represent national treasures and essential elements of the biosphere on which we all depend.

**References**

Binswanger, H., and P. Pingali. 1984. "Population Density and Agricultural Intensification: A Study of the Evolution of Technologies in Tropical Agriculture." Washington, D.C.: World Bank, Agriculture and Rural Development Department.

Boado, E. L. 1988. "Incentive Policies and Forest Use in the Philippines." In *Public Policies and the Misuse of Forest Resources*, ed. R. Repetto and M. Gillis. 165–203. New York: Cambridge University Press.

Boserup, E. 1981. *Population and Technological Changes*. Chicago: University of Chicago Press.

Burgess, P. F. 1988. "Natural Forest Management for Sustainable Timber Production: The Asia Pacific Region." Report, August 1988. Yokohama, Japan: International Institute for Environment and Development and International Tropical Timber Organization.

Chang, S. J. 1981. "Determination of the Optimal Growing Stock and Cutting Cycle for an Uneven-aged Stand." *Forest Sciences* 27(4):739–744.

Detwiler, R. P., and C. A. S. Hall. 1988. "Tropical Forests and the Global Carbon Cycle." *Science* 239:42–47.

Hadley, M. 1988. "Rain Forest Regeneration and Management: Report of a Workshop." Guri, Venezuela, 24–28 November 1986, *Biology International*, Special Issue 18, 1988.

Hartman, R. 1976. "The Harvesting Decision When the Standing Forest Has Value." *Economic Inquiry* 14(1):52–58.

Harvard Institute for International Development. 1988. *The Case for Multiple Use Management of Tropical Hardwood Forests.* Prepared for the International Tropical Timber Organization, Yokohama, Japan, January.

HIID. See Harvard Institute for International Development.

Houghton, R. A., R. D. Boone, J. R. Fruci, J. E. Hobbie, J. M. Melillo, C. A. Palm, B. J. Peterson, G. R. Shaver, G. M. Woodwell, B. Moore, D. L. Skole, and N. Myers. 1987. "The Flux of Carbon from Terrestrial Ecosystems to the Atmosphere in 1980 Due to Changes in Land Use: Geographic Distribution of the Global Flux." *Tellus* 39B:122–139.

Jodha, N. S. 1990. "Sustainable Agriculture in Fragile Resource Zones: Technological Imperatives." Discussion Paper Series, International Centre for Integrated Mountain Development, Kathmandu, Nepal, February.

Lanly, J.-P. 1982. *Tropical Forest Resources.* Forestry Paper 30. Rome: Food and Agriculture Organization.

Malingreau, J.-P., and C. J. Tucker. 1988. "Large Scale Deforestation in the Southeastern Amazon Basin of Brazil." *Ambio* 17(1):49–55.

Mellor, J. 1976. *The New Economics of Growth.* Ithaca, N.Y.: Cornell University Press.

Mergen, F., and J. Vincent, eds. 1987. *Natural Management of Tropical Moist Forests.* New Haven: Yale University School of Forestry and Environmental Studies.

Myers, N. 1988. "Threatened Biotas: 'Hotspots' in Tropical Forests." *Environmentalist* 8(3):1–20.

Nectoux, F., and Y. Kuroda. 1989. *Timber from the South Seas: An Analysis of Japan's Tropical Environmental Impact.* Zurich: World Wildlife Fund International Publication.

Peters, C. M., A. Gentry, and R. Mendelsohn. 1989. "Valuation of an Amazonian Rainforest." *Nature* 339:655–656.

Pfeffermann, G. Unpublished memorandum. "The Social Cost of Recession in Brazil." Washington, D.C.: World Bank.

Poore, D. 1988. *Natural Forest Management for Sustainable Timber Production.* Report for the International Tropical Timber Organization, Yokohama, Japan.

Prickett, G. T. 1988. *An Economic Comparison of Cattle Ranching and Extractive Production in Acre, Brazil.* New Haven: Yale University School of Forestry and Environmental Studies.

Ranis, G., and J. C. H. Fei. 1961. "A Theory of Economic Development." *American Economic Review* 51(4):533–565.

Repetto, R. 1988. *The Forests for the Trees? Government Policies and the Misuse of Forest Resources.* Washington, D.C.: World Resources Institute.

Repetto, R., and M. Gillis. 1988. *Public Policies and the Misuse of Forest Resources.* New York: Cambridge University Press.

Repetto, R., and T. Holmes. 1983. "The Role of Population in Resource Depletion." *Population and Development Review* 9(4):609–632.

Repetto, R., W. Magrath, M. Wells, C. Beer, and F. Rossini. 1989. *Wasting Assets: Natural Resources in the National Income Accounts*. Washington, D.C.: World Resources Institute.

Rietbergen, S. 1988. "Natural Forest Management for Sustainable Timber Production. Volume 2: Africa." London: Regional report from International Institute for Environment and Development for the International Tropical Timber Organization, October.

Salati, E., and P. G. Vose. 1983. "Depletion of Tropical Rain Forests." *Ambio* 12:67–71.

Schmidt, R. C. 1989. "Management of Tropical Moist Forests in Brazil." Report for the government of Brazil by the Food and Agriculture Organization of the United Nations, April.

Setzer, A. W., et al. 1988. "Relatorio de actividas do projeto IBDF-INPE 'SEQE'— ano 1987." Instituto de Pesquisas Espacias, São José dos Campos, Brazil, May.

Woodwell, G. M., J. E. Hobbie, R. A. Houghton, J. M. Mellilo, B. Moore, B. J. Peterson, and G. R. Shaver. 1983. "Global Deforestation: Contribution to Atmospheric Carbon Dioxide." *Science* 222:1081–1086.

World Bank. 1988. *Forest Resource Management Project: Ghana*. Report of the Western Africa Department of the World Bank, Washington, D.C., 17 November.

# 6

# Conservation and Sustainable Development of Forests Globally: Issues and Opportunities

## Jagmohan S. Maini and Ola Ullsten

Forests have emerged as one of the priority items on the international policy agenda, particularly in the context of the United Nations Conference on Environment and Development held in Brazil in June 1992. While special interest groups are focusing on a specific role, service, or value of forests (for example, as a source of biodiversity or a reservoir for carbon), the national and international scientific and policymaking communities face the challenge of reconciling national economic and environmental policy objectives for forests with the global environmental interests of the community of nations. This chapter examines the broad historic, environmental, industrial, social, cultural, and geopolitical contexts surrounding the conservation and sustainable development of all types of forests worldwide and offers potential policy responses at local, national, and international levels.

**Context**

*Historic.* During the past ten thousand years, global forest cover has been reduced by about one third, from an estimated $6.3 \times 10^9$ ha to about $4.2 \times 10^9$ ha. A considerable proportion of the historic deforestation has taken place in the temperate and boreal regions to meet the needs of an expanding population. Recent FAO (1988) data indicate that seventeen countries account for about 75 percent of the global forest cover, and, of those seventeen, five countries, namely Russia, Brazil, Canada, the United States, and Zaire, account for nearly 55 percent. According to Allan and Lanly (1991) boreal forest cover has stabilized and temperate forest cover is expanding slightly, but tropical forest cover shrank in the late 1980s at the rate of $17 \times 10^6$ ha per year. The tropical forests are now under pressures similar to those experienced by the temperate and boreal regions during the past few centuries.

*Environmental.* The environmental services of forests such as soil and water conservation; offering a habitat for diverse flora and fauna and a rich reservoir of biodiversity; and participation in ecological cycles (for example, carbon, oxygen, nutrient, hydrologic, and climatic cycles) are receiving increasing consideration by the scientific and policymaking communities at both national and international levels. The role of forests in the biodiversity and climate-change conventions, negotiated in association with the UNCED process, is receiving particular attention. From a policy perspective, the international discussions tend to be polarized along four main lines.

First, the industrialized countries, which are responsible for a significant proportion of global environmental problems, such as warming and atmospheric pollution, are advocating strong measures to conserve and protect the world's forests. The developing countries are concerned that the industrialized countries' preoccupation with tropical deforestation is inconsistent with the amount of attention being paid to prevent global warming and forest decline in Europe. They see here an equally urgent need to minimize harmful emissions and reduce the use of fossil fuels by the North. It should also be noted that deforestation is not the principal cause of the anticipated global warming, nor is reforestation the principal solution. Large-scale reforestation programs, including rehabilitation of derelict lands, would provide multiple environmental, economic, social, and cultural benefits, however.

Second, many developing countries view attempts to protect and preserve tropical forests as "locking up" their forests, constituting an intrusion into their sovereign rights to develop their resources in order to meet national policy objectives. Some also view this as an attempt by industrialized countries to impede the economic development of the South. Third, the commitment of developing countries to conserve the world's forests and biodiversity and to practice sustainable development remains conditional on obtaining from the richer, industrialized countries additional funding and technologies at favorable terms. The developing countries seek compensation for the economic opportunities foregone.

And finally, many developing countries are concerned about the desire of some industrialized countries to have free access to the genetic resources in the tropical forests and to the traditional knowledge of the indigenous inhabitants. The developing nations expect the North to pay for these precious genetic resources and traditional knowledge, just as they are expected to pay for technologies from the industrialized nations.

*Industrial.* The rate of increase in the demand for forest products usually exceeds the rate of population increase. Current projections suggest that over the next three decades, the increase in demand for forest products is expected to be about 3 percent per year. A number of developing countries that were formerly net exporters of forest products are now net importers. This trend is increasing. Economists and environmentalists attribute this unfortunate situation to the need of developing countries to liquidate forest assets in order to generate the capital required for economic development and to pay foreign debt. The lack of adequate funds to invest in reforestation, forest management, and forest protection is resulting in a net loss in forest cover in many parts of the world (Maini 1990a). The world forestry community faces the challenge of meeting increasing demand for a wide range of industrial and nonindustrial forest products from a shrinking resource base (Maini 1991b).

"Green consumerism" is another factor emerging in international trade in forest products. Many consumers in industrialized countries are choosing to buy forest products from sustainably managed forests manufactured by environmentally sound processes and technologies. Many consumers, individually and collectively (like municipal governments), have threatened to boycott tropical hardwood products. In order to protect their international trade, countries exporting forest products will have to formulate internationally recognized criteria for sustainable forest development (Maini 1991b) and apply these criteria to domestic forestry practices.

*Social and Cultural.* Forests play a wide-ranging social and cultural role in various parts of the world. Following the publication of the Brundtland report (World Commission on Environment and Development 1987), conservation and sustainable development for the present and future generations (intergenerational responsibility) has emerged as a global ethical issue, which is actively discussed, particularly in the industrialized countries (Maini 1990b). This issue, however, attracts limited attention in developing countries, where today's survival is the major preoccupation.

Closely related to the ethical issue of intergenerational responsibility is the need to protect the rights of forest dwellers, indigenous people and communities living in and around forests who are also dependent on them (Maini 1991a; see also chap. 3). Interest grows in the use of indigenous or traditional knowledge for the conservation and sustainable development of forests. Issues associated with the transfer of traditional knowledge, such as intellectual property rights and financial compensation, are now emerging in international deliberations.

*Geopolitical.* Forests represent a unique problem in terms of global environmental issues. Physically, they are located within the territories of sovereign nations. Yet their environmental role extends well beyond their borders (Maini 1991b). Management or mismanagement of watershed forests of international rivers, for example, offers transboundary implications in terms of soil and water conservation in neighboring countries. Similarly, airborne pollutants generated in one country may be transported into neighboring countries and cause forest decline (Ullsten 1991a). The role of forests in global carbon, oxygen, and climatic cycles emphasizes their environmental significance far beyond the boundaries of the nations where they are located. In this context, all types of forests worldwide are being viewed as similar to global commons like the atmosphere and oceans.

The national and international policymaking communities are faced with the need to reconcile national interests with international responsibilities and to develop appropriate policy responses, instruments, and institutions to accomplish this balance. Conservation and sustainable development of the global forests falls firmly on the international policy agenda. But this issue initially emerged largely from a narrow environmental and protectionist perspective. The protection of forests was sought because of their environmental value as sources of biodiversity and carbon reservoirs. This important but limited perspective must enter the broad spectrum of potential values and benefits to be derived from the forests at local, national, and international levels.

We must now enlarge the debate, from "combating deforestation" to "the conservation and sustainable development of all types of forests worldwide." This crucial change provides a more cohesive and comprehensive framework with which to address a wide range of issues *and opportunities* associated with forests (Maini 1991b). The world forests first and foremost represent a renewable economic resource of immense importance for any country's economic development. Any nonsustainable use of forests thus means economic loss. For many developing countries, where the destruction of forests is taking place at an alarming pace, deforestation is indeed an economic disaster. Any formulation of the global forest issue, therefore, has to recognize that for the developing countries, the economic rather than the environmental aspects of forest management form a priority. We do not intend to understate the significance of ecological considerations. But forests will benefit the environment as well as the economy if they are preserved.

Forests should be viewed as a source of a wide range of socioeconomic and environmental outputs and services. It is in the collective interest of the community of nations to cooperate in formulating a world-forests strategy in the broad context of the role that forests and forest products play in meeting basic human needs (Maini 1991a). From the point of view of the international forestry community the relevant questions surrounding the sustainable development of forests worldwide should include: Given the expanding world population and an otherwise triggered increase in the demand for wood products, how do we meet that demand without a devastating exhaustion of the forest resource base? and which technical, financial, and political means are available to practice and promote sustainable forest management in tropical countries? Formulation of these questions then brings us face to face with the complexity of the underlying causes of deforestation and forest decline and points to the need to undertake collective and complementary actions at local, national, and international levels.

### Approaches to Sustainable Development of Global Forests

It is apparent that different factors influence sustainable forest development for the industrialized boreal and temperate regions than act for the developing tropical regions. The same applies to the success of cooperation between the donor and the recipient countries.

There is an increase in green consumerism in the industrialized countries; consumers prefer to buy forest products that are obtained from sustainably managed forests and manufactured with environmentally acceptable technologies. Threats of boycotts of tropical hardwood products by consumers and municipalities in several industrialized countries are likely to be counterproductive, however. Forests without economic, social, environmental, or other values are unlikely to be conserved or sustainably managed. Movements toward the conservation and sustainable management of forests are likely to succeed through profits, benefits, and incentives, not by punitive action.

The tropical and subtropical regions need urgent collective action in terms of funds, technology, and training by the community of nations to assist in social forestry, agroforestry, afforestation, reforestation, sustainable forest development, forest biodiversity conservation, and rehabilitation of degraded lands. Planting individual trees or afforestation programs, as well as conservation and sustainable development of forests, will succeed only if these activities benefit local communities and national interests. No one in the developing countries will plant trees only to create carbon reservoirs that will allow citizens in industrialized countries to drive their cars and run their factories. Although forests in the boreal and temperate regions are generally managed and used in a sustainable manner, reduction of airborne pollutants associated with forest decline remains a major issue that needs to be addressed. At the national level conservation and sustainable development of forests must be on the political agenda and demonstrated by clearly enunciated policies that are supported by effective national programs and institutional arrangements (Maini 1991a). This factor is critical to the success of international cooperative efforts. At the international level, governments must establish an effective policy and institutional framework to facilitate and coordinate international cooperation on the conservation and sustainable development of forests.

Communities must plant trees and forests to provide immediate benefits and meet local needs for food, fiber, fodder, shelter, and economic development. Local individuals and communities should be involved in developing and implementing action plans. They must have "ownership" of these initiatives. The basic causes of deforestation and forest degradation should be addressed by, for example, providing alternate resources.

National land-use policies that promote conservation and sustainable development of forests should be developed and implemented, as should forest policies on permanent forest cover that have clear targets and timetables for their achievement. We would propose three types of targets, moving from the most to the least extensive. The first target could be total forest cover, embracing all kinds of forests—natural, replanted, and new plantations. It would involve planting programs and the rehabilitation of degraded lands.

The second target would be permanent forest cover, embracing those areas of a country which have been designated through national land-use policies to be permanently kept under forest cover, rather than left open for conversion to other kinds of land use. A number of delegations to UNCED have already referred to the need for this action in their interventions on Guiding Principles on Forests. This target would include forest land slated for commercial exploitation under management regimes that permit regeneration of the forest, in order to maintain its status as productive forest land.

The third and the least extensive target would be protected natural forests, embracing examples of representative and unique forest types protected from commercial exploitation in order to conserve their biodiversity and other ecological values.

A clear institutional focus on the national level should be developed with the responsibility and accountability for forest-related policies and programs. Government agencies should produce timely and reliable reports on the state of forests at national levels. These reports would include economic, environmental, social, and cultural dimensions of forests; they would not merely offer a traditional forest inventory with an industrial-use perspective on forests. Governments must also invest in forest research, particularly in the management of forest ecosystems for multiple values and in the rehabilitation of degraded forest lands. In addition promoting exports of forest products from sustainably managed forests needs urgent attention.

On the global level, developing an international consensus on the guiding principles applicable to the conservation and sustainable development of all types of forests worldwide represents a first step toward progress (Maini 1991a). It is not enough, however. The formulation of these guiding principles should lead their framers to organize an international convention on forests, which will provide the policy and strengthened institutional framework for international cooperation.

Strong international institutions are needed to provide leadership on forest-related issues, to manage international cooperation on forests, to assess the state of world's forests, and to provide timely and credible reports on the progress toward the conservation and sustainable development of forests at regional and global levels. Enhanced regional cooperation, particularly among contiguous states sharing a watershed, a particular forest type, or a unique forest ecosystem, offers an important complement to international cooperation.

The interactions among trade, environmental degradation, and economic development have recently become the subject of much scrutiny. Trade of forest products is perhaps one of the more salient topics within that issue. International trade in forest products from sustainably managed forests should be promoted. Yet a necessary prerequisite is the formulation of internationally accepted criteria for sustainable forest development.

The plan we propose, "Project Green Globe," offers an important opportunity for the international community to progress toward these forest-management goals. The international community must establish realistic targets for increasing global forest cover, to be achieved by the years 2000, 2010, and 2025. Appropriate funding and cooperation in sharing technology and other kinds of knowledge that will help accomplish targets identified under "Project Green Globe" must also be arranged. Increasing global forest cover is important, but it is also crucial to control transboundary airborne pollutants that cause forest decline.

Although conservation and sustainable development of all types of forests worldwide is in the collective interest of the community of nations, the primary responsibility for formulating and implementing appropriate policies and practices lies with national governments. The establishment and implementation of strong national forest policies and a clear institutional focus for forest-related activities would permit the international community to assist individual nations to meet nationally established targets and timetables. Our collective targets and timetables could be mobilized under the proposed "Project Green Globe." Under it, a strong international institutional focus would allow for coordinated international efforts and monitoring of progress.

Conservation and sustainable development of forests will be achieved through profits and incentives at the community and national levels, not by punitive actions. Consequently, international trade in forest products from sustainably managed forest should be promoted.

The international community is making significant progress in developing a global consensus on the guiding principles for the conservation and sustainable development of all types of forests worldwide, within the UNCED process. An international convention on all types of forests worldwide marks the next logical step. It would provide an international policy and institutional framework to strengthen multi- and international cooperation and national policies and institutions dealing with forest-related issues and opportunities.

**References**

The views expressed in this chapter are personal views of the authors and do not necessarily reflect official positions of their organizations or governments.

Allan, T., and J. P. Lanly. 1991. "Overview of Status and Trends of World Forests." In *Proceedings of the Technical Workshop to Explore Options for Global Forestry Management,* ed. D. Hewlett and C. Sargent, 17–39. Bangkok, 24–30 April 1991. London: International Institute for Environment and Development.

FAO. 1988. *An Interim Report on the State of Forest Resources in the Developing Countries.* Rome: Food and Agriculture Organization.

Maini, J. S. 1990a. "Forests: Barometers of Environment and Economy." In *Planet under Stress,* ed. C. Mungall and C. D. McLearen, 168–187. Oxford: Oxford University Press.

———. 1990b. "Sustainable Development in the Canadian Forest Sector." *Forestry Chronicle* 66(4):346–349.

———. 1991a. *Guiding Principles Towards a Global Consensus for the Conservation and Sustainable Development of All Types of Forests World-wide.* Ottawa: Department of Forestry.

———. 1991b. "Towards an International Instrument on Forests." In *Proceedings of the Technical Workshop to Explore Options for Global Forestry Management,* ed. D. Hewlett and C. Sargent, 278–285. Bangkok, 24–30 April 1991. London: International Institute for Environment and Development.

Ullsten, O. 1991a. Foreword to *European Forest Decline: The Effects of Air Pollutants and Suggested Remedial Policies,* ed. S. Nilsson. Laxenburg, Austria: International Institute for Applied Systems Analysis.

———. 1991b. "Keynote Speech." In *Proceedings of the Technical Workshop to Explore Options for Global Forestry Management,* ed. D. Hewlett and C. Sargent, 10–13. Bangkok, 24–30 April 1991. London: International Institute for Environment and Development.

World Commission on Environment and Development. 1987. *Our Common Future* (the Brundtland report). Oxford: Oxford University Press.

# 7

# The Need for an International Commission on the Conservation and Use of World Forests

**Kilaparti Ramakrishna**

The inability of nations to manage their forest resources is evident in both the high rates of deforestation and the rapid depletion of forest biodiversity. As concerns about global warming grow, the role that forests play in creating global carbon sinks has received significant attention in policymaking circles. Forests store approximately three times the amount of carbon currently contained in the atmosphere. Estimates suggest that deforestation is contributing $1-3 \times 10^9$ tons of carbon per year as carbon dioxide to the atmosphere. Although combustion of fossil fuels adds approximately $5.6 \times 10^9$ tons per year and is thought to be a major contributor to the approximately $4 \times 10^9$ tons of carbon that accumulate each year in the atmosphere (Houghton 1990; Houghton et al. 1985, 1987), I shall here emphasize only the role that forests play in the carbon-flux equation.

The Intergovernmental Panel on Climate Change (IPCC), set up jointly by the World Meteorological Organization (WMO) and the United Nations Environment Programme (UNEP), pointed out in its first assessment report (Houghton et al. 1990) that carbon dioxide is a principal greenhouse gas and that increases of greenhouse gases will raise the global temperature. Several scientists believe that warming will stimulate respiration and speed the release of still additional carbon dioxide from currently intact forests (Woodwell 1983, 1985, 1988). This release might add several billion tons annually to the amount of carbon currently released from deforestation. Forest-induced carbon dioxide increases in the atmosphere come on the one hand from deforestation, driven by economic and political pressures, and on the other from a net further release of carbon dioxide from increasing rates of respiration as the earth warms.

Many of the solutions suggested for the threat of climate change call for stopping deforestation and shifting globally to reforestation as rapidly as possible. Questions arise, however, as to how to bring about such changes. Unless specific plans can be provided, complete with details of the local, regional, national, and global advantages of compliance, there will be no progress. Even to think about such plans we need to answer basic questions: Are forests necessary? If so, how many and of what kind, where and for what? This book has asked and, we hope, answered many of these. It has also tried to investigate the driving forces behind deforestation, consider some of the remedial measures attempted, and recommend plans for monitoring their success at local, regional, national, and global levels. Without satisfactory answers to or at least awareness of these questions, the fact that there is a global consensus for the protection and preservation of world forests means little, and the attempts to create an international agreement on the issue appears bound to fail.

### The Idea of Regulating Forests Globally

Several recent studies have indicated that international agreements for the protection of the environment have been increasing since 1972 (Kiss 1983; Ruster and Simma 1975–82; UNEP 1991; Scovazzi and Treves

1992; ECE 1992; see these sources as well for the treaties and conventions discussed below). Yet few, if any, deal specifically with forest protection. The earliest reference to forest protection in an international agreement is found in the *Convention Relative to the Preservation of Flora and Fauna in their Natural State, 1933*. This was adopted in London and it entered into force in 1936. Its objectives are to preserve the natural flora and fauna of certain parts of the world, particularly Africa, by establishing national parks and reserves and regulating hunting and species collection. Articles 3 and 4 of the convention deal with the establishment of natural parks and strict natural reserves, as well as the control of all human settlements in those areas. This convention was open for accession to all nations but so far only nine countries have become parties to it: Belgium, Egypt, India, Italy, Portugal, South Africa, the Sudan, the United Kingdom, and the United Republic of Tanzania.

The next convention related to forests is the *International Convention for the Protection of Birds*, which was adopted in Paris in 1950, to become effective in 1963. The delegates to this conference were also interested in establishing reserves, this time to provide breeding grounds for birds. The objective stated by the convention is "to protect birds in the wild state, considering that in the interests of science, the protection of nature and the economy of each nation, all birds should as a matter of principle be protected." This convention has only ten countries as parties, however: Belgium, Iceland, Italy, Luxembourg, the Netherlands, Spain, Sweden, Switzerland, Turkey, and Yugoslavia.

Among the United Nations and its specialized agencies the Food and Agriculture Organization, established 1945, has the primary mandate in forestry matters. The constitution of FAO calls for the organization to promote and where appropriate to recommend national and international action with respect to the conservation of natural resources and the adoption of improved methods of forest production. Other organizations whose interests spill into this area include the United Nations Environment Programme (UNEP), the United Nations Development Programme (UNDP), the United Nations Conference on Trade and Development, the United Nations Industrial Development Organization, and the World Bank. Yet growing recognition of the unique role forests play globally suggests that we now need an international organization designed specifically to deal with forests.

Among regional agreements on the subject of forests, the following merit special attention. The *Convention on Nature Protection and Wildlife Preservation in the Western Hemisphere*, adopted in 1940, entered into force in 1942. El Salvador, Guatamala, Haiti, the United

States, and Venezuela were the initial parties, and they have since been joined by Argentina, Brazil, Chile, Panama, Peru, Trinidad and Tobago, and Uruguay. The objective of the convention is to preserve all species and genera of native American fauna and flora from extinction and to preserve areas of extraordinary beauty, striking geological formation, or aesthetic, historic, or scientific value. Article 2 obliges parties to establish national parks and reserves, nature monuments, and strict wilderness reserves, while article 4 asks that the strict wilderness areas be maintained inviolate.

*The Convention on the Conservation of Nature in the South Pacific, 1976*, which took effect in 1990, has for its parties six countries: Australia, the Cook Islands, Fiji, France, Papua New Guinea, and Western Samoa. The convention's objective is to take action for the conservation, utilization, and development of the natural resources of the South Pacific region. Articles 2–5 touch upon areas of interest to forest conservationists. Article 2 enjoins parties to create protected areas to safeguard representative samples of natural ecosystems. Article 3 prohibits national parks from being altered so as to reduce their area except after the most complete investigation, and article 4 provides that national reserves be maintained inviolate, as far as is practicable.

Another agreement governing the same region is the *Convention for the Protection of Natural Resources and Environment of the South Pacific Region, 1986*. It took effect in 1990. Its objective is to protect and manage the natural resources and environment of the South Pacific region. Article 14 in particular provides that all appropriate measures to protect and preserve rare ecosystems and endangered flora and fauna as well as their habitat must be taken. Parties to this convention number ten: Australia, the Cook Islands, the Federated States of Micronesia, Fiji, France, the Marshall Islands, New Zealand, Papua New Guinea, the Solomon Islands, and Western Samoa.

*The Treaty for Amazonian Cooperation*, adopted in 1978, entered into force in 1980. Bolivia, Brazil, Colombia, Ecuador, Guyana, Peru, Suriname, and Venezuela were its parties, and through it they wished to promote the harmonious development of the Amazonian region and permit equitable distribution of the benefits of such development among themselves. Although the treaty provides for the parties to undertake joint efforts to promote the development of their Amazonian territories, to preserve the environment, and to foster the conservation and rational use of natural resources, its primary purpose is to stress international cooperation and recognition of the exclusive sovereign right of each country to use the natural resources within the territory of each party. In addition the treaty acknowledges the complete freedom of the Amazon and other international Amazonian rivers for commercial navigation. The Amazonian Cooperative Council, created under the treaty, calls for top-level diplomatic representatives to meet once a year to supervise its effective implementation, under the general guidance of each country's Minister of Foreign Affairs.

Another regional effort of a similar nature was a convention adopted by the Council of Europe called the *Convention on the Conservation of European Wildlife and Natural Habitats*. Agreed upon in 1979, it took effect in 1982, with Austria, Belgium, Burkina Faso, Cyprus, Denmark, Finland, France, Greece, Hungary, Ireland, Italy, Liechtenstein, Luxembourg, the Netherlands, Norway, Portugal, Senegal, Spain, Sweden, Switzerland, Turkey, the United Kingdom, and the European Community as parties. Its objective is to conserve wild fauna and flora and their natural habitats. The convention provides for the creation of nature reserves insofar as is necessary to achieve its objectives.

The *Benelux Convention on Nature Conservation and Landscape Protection* was adopted in 1982, and it entered into force in 1983. It has the Benelux countries (Belgium, Netherlands, Luxembourg) as its parties. In order to preserve nature, natural areas, and the landscape, especially in boundary regions, it provides for the protection of transboundary natural areas and landscapes. It also seeks to establish programs for the protection of such areas.

The *ASEAN Agreement on the Conservation of Nature and Natural Resources* was adopted in 1985 by the Association of Southeast Asian Nations (Brunei Darussalam, Indonesia, Malaysia, the Philippines, Singapore, and Thailand). Its objective is to promote joint and individual state action for the conservation and management of the natural resources of the ASEAN region. This is more comprehensive than other regional agreements and provides greater detail of how its objective is to be accomplished. The parties agreed to preserve genetic diversity by ensuring conservation and preservation of all species within their jurisdictions, to maintain harvested species through sound management and sustainable use, to promote soil conservation, improvement, and rehabilitation, to conserve ecological processes by reducing, controlling, or preventing environmental degradation and pollution, to set up protected areas to conserve biological diversity, especially of endangered species, to ensure that conservation and management of natural resources becomes an integral part of development planning at both national and regional levels, and finally to harmonize the use of shared resources without prejudice to the environment, avoiding harmful international environmental effects. The agreement is still to come into effect. But it was signed by all the member countries of ASEAN.

Although all of the agreements referred to above touch upon forests, the principal purposes of entering into them were not conservation and management of forest resources. The only international agreement to date that specifically provides for a framework governing a certain type of forest, even though principally for the purposes of cooperation and consultation between contracting parties, is the *International Tropical Timber Agreement* adopted in 1983, which entered into force in 1985. Forty-eight countries have joined so far as parties. The agreement provided for the establishment of the International Tropical Timber Organization, which now administers the provisions of the treaty and supervises its operation. The ITTO functions through the International Tropical Timber Council, also established under this agreement. The council arranges for consultations within the United Nations and with the special U.N. agencies and nongovernmental organizations. In addition the council established the Committee on Economic Information and Market Intelligence; the Committee on Reforestation and Forest Management; and the Committee on Forest Industry.

Most of the provisions of the agreement are tailored to increase international cooperation by facilitating consultations between countries producing and consuming tropical timber. In addition, its objectives include promoting research and development to improve forest management and wood use, encouraging national policies aimed at sustainable use and conservation of tropical forests and their genetic resources, and maintaining the ecological balance in the regions concerned.

The Tropical Forestry Action Plan (TFAP) was launched in 1985 by the World Bank, the UNDP, FAO, and the World Resources Institute as an emergency response to the tropical forest crisis. It calls for a massive increase in development assistance to the forestry sector. Under TFAP projects are developed as a result of national planning exercises, which involve a review of the problems facing the forest sector in the country concerned. The level of development assistance from the member countries has increased to $1 billion, up from $500 million per year before the adoption of the plan. But TFAP has not been able to deliver satisfactory results in most countries. Its critics believe rightly that its failure in large measure is due to its "top down" approach, which provides inadequate participation for people who live in and depend on tropical forests and for nongovernmental organizations. Given the character of the international system of states, this problem obviously raises important questions about what success even a global agreement can have in governing forests.

## A Comprehensive Global Agreement

When it was set up in 1988, the purpose of the IPCC was to submit a report bringing together all the available knowledge on the scientific analysis, socioeconomic effects, and policy responses necessary to address climate change. The process thus initiated was remarkable both in terms of what it was able to accomplish and how a scientific issue found its place on national agendas throughout the world.

In the time it took the IPCC to produce its first assessment report in August 1990 (Houghton et al. 1990), there have been innumerable meetings around the world on specific subsets of the issue of climate change. The flux of one of the major greenhouse gases, carbon dioxide, depends to a large extent on what nations do to the forests in their jurisdictions. Attention has quickly turned to the role forests can play as global carbon sinks and the need to conserve them for that reason (as well as for all the other purposes that have also not found a place in international responses dealing with forests).

At a workshop organized under the auspices of the IPCC and held in São Paulo during January 1990, the participants supported developing a world forest conservation protocol within the context of a climate-change convention. This initiative came about to ward off attention on tropical forests as a universal panacea for the carbon ills of the industrialized countries. It has since then gained a significant number of backers. Significant among them are the Report of the Independent Review Team of the Tropical Forestry Action Plan, headed by Ambassador Ola Ullsten, which called for a world forest convention. The report clearly points out that adopting an approach limited only to tropical forests offers no long-term solution to the problem of forest loss. The G-7 Summit in Houston in July 1990 also declared its readiness to negotiate a global convention necessary to arrest forest destruction and stimulate positive action to fight the threats facing forests throughout the world. Through FAO's committee on forests and its council, FAO in September and November 1990 approved the idea of an international instrument for the conservation and sustainable development of forests.

The World Forestry Congress has traditionally been a forum for promoting the timber industry's interests. The Tenth World Forestry Congress, however, held in Paris in September 1991 (see Appendix 1), realized that the wide-ranging threats to forests had become a concern of the entire international community. Recognizing that the devastating effects of deforestation and forest deteriorization ignore national frontiers, the Congress chose as its theme "Forest: A Heritage for the Future." Its final declaration hoped to establish a consensus of opinion and formulate recommendations for national forest policies and to influence discussions on the adoption of the statement of principles on forests for UNCED. Before the Earth Summit and the adoption of Agenda 21, the Paris Declaration was the most significant statement on forests as such. But it is still not enough.

When UNCED began its preparations for the 1992 conference in Rio de Janeiro, it was believed that the Preparatory Committee (PrepCom) should be in a position to examine all steps and options for a global consensus on the management, conservation, and development of all types of forest, either as an integral part of the Earth Charter (which was later renamed the Rio Declaration on Environment and Development) or as a separate item. Some interpreted this to mean that they should arrange for the negotiation and adoption of an international agreement governing forests. Given the emotional nature of the discussions on the topic the PrepCom quickly decided that the best course to pursue would be not to negotiate a convention for the conservation and use of forests but to adopt a set of guiding principles for a consensus on forests. The PrepCom produced a text at the end of their meeting in Geneva in September 1990 entitled "A non-legally binding authoritative statement of principles for a global consensus on the management, conservation and sustainable development of all types of forests." The only document ready for adoption at the 1992 UNCED conference was this statement concerning forests (see Appendix 2).

Several countries within the European community as well as outside it hope that these principles will lead to a framework convention on forests after the 1992 UNCED conference. Some within the European community are trying to provide a timetable for the negotiations of a convention on forests and hope that these negotiations will begin immediately following the conference. (At the Earth Summit, Prime Minister Felipe Gonzalez of Spain called for a world conference on forests and offered Spain as the host country.)

The nongovernmental organizations in the United States meeting under the umbrella Forest/Climate Working Group expressed four main concerns, many of which go beyond the ambit of governmental discussions, in terms of both scope and specificity of actors. They are expressed in the following recommendations:

• Halt the destruction of forests, both primary and habitats of biological importance in secondary temperate and tropical forests. This involves forest conservation and management strategies to maintain the long-term viability of forest ecosystems.

• Create an agreement that uses language addressing local issues important to the indigenous population and that includes the involvement of these people in protection strategies.

• Do not allow the agreement to serve as a substitute for the responsibility of industrialized countries to reduce greenhouse gas emissions, nor to side-step the need to reform existing international mechanisms, like TFAP, which have failed to halt the destruction of tropical forests. At the same time, a climate convention, independent of a free-standing forest convention, should include a forest protocol computable with the above concerns that extends beyond afforestation and reforestation efforts. It should be recognized that afforestation in certain circumstances can be damaging to the environment, and to avoid potential adverse effects quasi-natural forests should be established where possible.

• Insist that any delegation to negotiate an agreement reflect interdisciplinary expertise. The Forest/Climate Working Group has serious reservations about a forest agreement primarily controlled by foresters, and considers it essential that such an agreement include linkages between the many sectors and institutions affecting forests.

### National Management of Forests: The Example of India

A survey of municipal legislations indicates that forests have long been the subject of important laws and regulations at the municipal, regional, and national levels. Historically, municipalities and states have recognized the interest of the public in retaining forests intact. The reasons have been diverse: forests have served as a source of firewood, protected water supplies, helped control floods, supplied timber for construction, and provided for future needs for land or other resources associated with forested land.

New reasons for protecting forests have gradually become apparent. The rights of indigenous populations, the need to preserve biological diversity, and the role forests play as global carbon sinks offer a few more examples of the value of this natural resource. Yet in general the inter-

ests which stress the long-term benefits of forests, however well articu-
lated, have not been handled effectively. Although it is dangerous to
offer simplistic explanations for this, a cursory look at the way forests
have been managed in India will, it is hoped, help us recognize some of
the problems involved.

The Indian constitution, in effect since 1950, provides for a federal
republic with a parliamentary form of government. The constitution
sets forth detailed lists enumerating the areas of legislative jurisdiction
of the union (list I: the Union List) and the states (list II: the State List),
as well as areas of concurrent jurisdiction (list III: the Concurrent List).
The Union List and the State List give the parliament and the state
legislature exclusive jurisdiction over the entries contained in those
lists. Under the Concurrent List both the union government and the
states may regulate. A union law regarding a concurrent subject generally prevails over a state law on the same subject.

During the framing of the constitution (1946–49), the central issue
debated by the Constituent Assembly was how to allocate power between the national and state governments. Debate over where to locate
authority to regulate environmental matters was primarily a manifesta-
tion of this more fundamental power dispute. Pro-center members,
those who advocated giving the power to a centralized federal government, argued vehemently for retaining topics such as forests and agriculture on the Concurrent List, so that both the central and state governments could have an equal role in their management. Proponents of
more decentralized federalism, on the other hand, feared the coercive
effect on the states of giving the central government too much power
over such topics (Constituent Assembly of India 1949).

The Ministry of Agriculture had proposed amendments that would
have placed "coordination of the development of agriculture, including
animal husbandry, forestry and fisheries" on the Union List, and recla-
mation of waste lands, forest laws, and inland fisheries and fishery laws
on the Concurrent List. In support of these amendments, the Ministry
of Agriculture argued that the "forests have a great bearing on the general agricultural development and prosperity of the country as a whole"
and that "it is essential to ensure that no Province or State follows, even
inadvertently, a policy which will be detrimental to the rest of the
country" (Constituent Assembly of India 1949). These amendments,
however, were rejected.

Govind Ballabh Pant of the Constituent Assembly opposed locating
"forests" on the Concurrent rather than the State list. In response to
Prime Minister Jawaharlal Nehru's query as to what would happen if the

union (central government) adopted legislation relating to forests under its authority to undertake national planning, Pant replied: "If it is hoped that the Provinces can be made to cooperate against their own will by means of a central legislation, that hope is not likely to materialize" (Constituent Assembly of India 1949).

When the constitution of India was adopted, forests and the wildlife were placed under the legislative authority of the state governments, the constituent units of the Indian federation. This notwithstanding the federal government from the beginning had expressed interest in these topics and pursued the same actively through a variety of policies.

In India wildlife protection and forest management are treated together. The national forest policy (as opposed to national forest legislation), declared in 1952, emphasizes the inclusion of wildlife protection within the field of forest management. The Indian Wild Life Protection Act of 1972 (passed at the request of eleven of the state governments and originally applying only to them; later it was adopted by the rest of the country) likewise treats forests and wildlife together and provides for the establishment of national parks.

Administrative organization in the central government also combines wildlife conservation and forest management. The Forestry Division, within the Ministry of Agriculture, is headed by an inspector general of forests who is also the chief executive of forestry and wildlife, and a joint secretary is responsible for forests and wildlife.

Because of poor and indifferent management practices, particularly with regard to forests, the Indian Parliament by a constitutional amendment act transferred forest and wildlife management from the exclusive purview of states to an area where both the state and central governments could exercise legislative control. This new situation gave the central government the power to act directly to protect forests and wildlife. In pursuance of that power the Indian Parliament in 1980 enacted the Forest Conservation Act, without the prior approval of the states. This act prevents states from removing land from reserved forest status without first getting the approval of the central government. The act also prohibits states from allowing any breaking up or clearing of forest land for any purpose other than reforestation without central government approval.

India is a member of the International Co-ordinating Council of the Programme for Man and the Biosphere, a worldwide program designed to study the structure and functions of ecosystems and the impact on them of human intervention. The program seeks to conserve representative ecological areas known as "biosphere reserves" and to preserve genetic material contained in these regions.

In 1980 the government of India decided to create biosphere reserves. Draft legislation was prepared in cooperation with the International Union for Conservation of Nature and Natural Resources and the Indian government prepared comprehensive project documents for five reserves. Altogether India has identified twelve sites, some of which extend into more than one state, as potential biosphere reserves. But controversy over the roles of the state and central governments in administering the reserves has delayed their establishment.

Arguments for central government control over biosphere reserves are strong. First, many identified reserves are larger than established national parks and extend into more than one state, which makes central administration desirable. In addition, the biosphere reserve project participants recommend that the project be financed entirely by the central government.

The state governments have come forward with a set of their own arguments. The Indian government proposes to implement the biosphere reserves program by amending the Indian Wild Life Act. This act is administered primarily through state agencies. Though the biosphere reserve plans call for some portion of management by central government officials, a large number of the scientific, administrative, and field personnel will have to come from state administrations. In addition states are reluctant to yield control of biosphere reserves because the reserves can provide substantial revenues. Musk deer found in the Namdapha Biosphere Reserve, for example, could be farmed and their valuable musk excretions harvested.

The unresolved questions involving this particular biosphere program are typical of problems likely to face any global issue. Management of world forests, which involves local, national, regional, and international groups all vying for control, each with a unique agenda, seem certain to find these problems multiplied indefinitely.

Implicit in the ongoing discussion of forests is the assumption that there will be a global forest convention soon after the UNCED conference in 1992. The question, however, is whether that convention is now inevitable or whether more must be done to prepare countries in order to bring them to the negotiating table. The example of India illustrates that considerably more thought needs to be given even to national management of forests. The challenge looms more formidably on an international scale. The TFAP does not offer an example to be emulated.

Constituency building is necessary to give the issue systematic, detailed, and careful consideration by the scientific and political communities jointly over a period of months to years. Their first step will be to define the issues systematically as questions and to bring these questions to the attention of scholars. International progress in addressing issues of environment has never been achieved without a definition of the problem and an equally clear definition of the solution by the scientific community. The efforts to protect the ozone layer and to stabilize and reduce greenhouse gases, though weak, are slowly gaining in strength as the scientific community builds its consensus on the causes. Such an exercise necessarily needs to be undertaken not within the existing agencies but by a new, ad hoc, fixed-term body. A combination of efforts that shaped the IPCC and the World Commission on Environment and Development is required to establish the proposed International Commission on the Conservation and Use of World Forests.

## References

Constituent Assembly of India. 1949. *Constituent Assembly Debates: Official Reports.* 9:719–964A. New Delhi: Constituent Assembly of India.

ECE. 1992. *Environmental Conventions.* New York: Economic Commission for Europe, United Nations.

Houghton, J. T., G. J. Jenkins, and J. J. Ephraums, eds. 1990. *Climate Change: The IPCC Scientific Assessment.* Cambridge: Cambridge University Press.

Houghton, R. A. 1990. "The Future Role of Tropical Forests in Affecting the Carbon Dioxide Concentration of the Atmosphere." *Ambio* 19:204–209.

Houghton, R. A., R. D. Boone, J. M. Melillo, C. A. Palm, G. M. Woodwell, N. Myers, B. Moore, and D. L. Skole. 1985. "Net Flux of $CO_2$ from Tropical Forests in 1980." *Nature* 316:617–620.

Houghton, R. A., R. D. Boone, J. R. Fruci, J. E. Hobbie, J. M. Melillo, C. A. Palm, B. J. Peterson, G. R. Shaver, G. M. Woodwell, B. Moore, D. L. Skole, and N. Myers. 1987. "The Flux of Carbon from Terrestrial Ecosystems to the Atmosphere in 1980 Due to Changes in Land Use: Geographic Distribution of the Global Flux." *Tellus* 39B:122–139.

Kiss, A. C. 1983. *Selected Multilateral Treaties in the Field of Environment.* UNEP Reference Series 3. Nairobi: United Nations Environment Programme.

Ruster, B., and B. Simma, eds. 1975–82. *International Protection of the Environment,* vols. 1–30. Munich: Springer-Verlag.

Scovazzi T., and T. Treves. 1992. *World Treaties for the Protection of the Environment.* Milan: Istituto per L'Ambiente.

UNEP. 1991. *Register of International Treaties and Other Agreements in the Field of the Environment.* Nairobi: United Nations Environment Programme. UNEP/GC.16/Inf.4.

Woodwell, G. M. 1983. "Biotic Effects on the Concentration of Atmospheric Carbon Dioxide: A Review and Projection." in *Changing Climate,* National Academy of Sciences–National Research Council. Washington, D.C.: National Academy Press.

———. 1985. "On the Limits of Nature," in *The Global Possible: Resources, Development, and the New Century,* ed. Robert Repetto, 47–65. New Haven: Yale University Press.

———. 1988. "Rapid Global Warming: Worse with Neglect." Testimony before the Senate Committee on Energy and Natural Resources, 23 June.

The Tenth World Forestry Congress
having assembled more than 2,500 participants from 136 countries
from 17th to 26th of September 1991,

*considering* the theme of the eighth World Forestry Congress, held in
Jakarta in 1978, *"Forests for People"*;

*considering* the theme of the ninth World Forestry Congress, held in
Mexico City in 1985, *"Forest Resources in the Integral Development of
Society,"* and its manifesto which urged *"all human beings of all na-
tions and their governments, within the framework of their own sov-*

*ereignty, to recognize the importance of forest resources for the bio-sphere and the survival of humanity";*

*considering* the International Conference "SILVA," held in 1986, which concluded with the *"Proclamation of Paris on Trees and Forests"*;

*considering* its own general theme *"Forests, a Heritage for the Future"* and all its detailed conclusions and recommendations that it has adopted on each theme discussed;

*considering* the general concern about deforestation and degradation of the world's forests caused by competition for land, inadequate management and the emission of pollutants generated by human activities, all of which have caused, in various regions of the world, at different times, and to varying degrees of irreversibility, the deterioration of the forest heritage;

*considering* that, rather than forest exploitation, the real causes of deforestation in developing countries are poverty, debt, underdevelopment, and the requirement to meet the basic needs of rapidly growing populations;

*considering* that forest resources are an important factor of socio-economic development, and more especially of rural development;

*considering* the responsibility of our generation to future generations for the world's natural heritage;

ADDRESSES public opinion, political leaders, international, intergovernmental and non-governmental organizations, from the whole,

REMINDS THEM

—the importance of the renewable goods and services provided by trees and forests, in the face of growing demand for building materials, fuel, wildlife, food, fodder, recreation areas, . . . ;

—the wealth and diversity of world forests, and their positive role in water and carbon cycles, in the protection of soils and the conservation of biodiversity;

—the availability, often ignored, of methods of management of trees and forests which can sustain and even increase the amount of goods and services they provide;

—the need to avoid irreversible damage to the biosphere; and thus the need for long-term planning in the management of natural resources;

AFFIRMS

—that the real challenge is to reconcile the economic use of natural resources with the protection of the environment through an integrated and sustainable development approach;

—that the solution to forest problems requires combined efforts to reduce poverty, increase agricultural productivity, ensure food security and energy supplies, and promote development;

—that the very concept of forest management constitutes a real tool to manage their economic, ecological, social and cultural functions, thus broadening the notion of sustained yield;

—that the integral conservation of particular forests for the protection of biodiversity constitutes a management objective;

AND RECOMMENDS

—that communities be involved in the integrated management of their land and that they be provided the necessary institutional, technical and financial means;

—that the long-term use of lands be planned on the basis of their potentialities to determine those which are suitable for forestry; and, in so doing, that attention be paid to the needs of people concerned, particularly those who depend on forests for their livelihood;

—that the continuity of tree and forest management policies be ensured, given the length of forest cycles;

—that the designation of certain representative or endangered forests as protected areas be pursued and that they be organized in national or international networks;

—that appropriate silvicultural techniques, increased planting and the perennial use of wood be developed to contribute to the absorption of carbon dioxide;

—that development of agroforestry systems, afforestation and reforestation be intensified.

The Tenth World Forestry Congress
aware of the gravity, the urgency, and the universality of developmental and environmental problems, but emphasizing the renewable nature of forest resources, and convinced of the soundness of solutions afforded by sustainable management of the world's forests, within the framework of national forestry policies,

SOLEMNLY CALLS UPON DECISION-MAKERS TO

*commit* themselves to the greening of the World through afforestation, reforestation and sustainable management of the multiple functions of trees and forests in the form of integrated programmes, involving the participation of the concerned populations, in accordance with national land use planning policies;

*regularly assess* developments in the forest heritage at national and international levels, drawing on the *"1990 World Forest Resources Assessment"* carried out by FAO;

*limit* all emissions of pollutants that damage forests, and *contain* emissions of greenhouse gases, including those produced by power generation;

*adapt* economic and financial mechanisms to the long-term requirements of forest management, and *increase* national and international funding, particularly for developing countries;

*work* towards the harmonious development of international trade in forest products through the prohibition of any unilateral restriction not in conformity with GATT; and *promote* the utilization of forest products;

*develop* cooperation at the political level on forest issues of regional importance, such as desertification control, forest protection, management of major watersheds, etc.;

*strengthen* and *coordinate* research and experimentation, training, exchange of information, and cooperation in all disciplines related to sustainable management of forest ecosystems;

*strengthen* activities of, and coordination among, the relevant international organizations;

*integrate* its conclusions and recommendations into the process of the United Nations Conference on Environment and Development (UNCED), in order to define *"A non-legally binding authoritative statement of principles for a global consensus on the management, conservation and sustainable development of all types of forests"*; and, into the current negotiations on biodiversity and climate change being conducted under the auspices of the United Nations;

*strengthen* international cooperation, particularly in the framework of the Tropical Forests Action Programme [sic] (TFAP), of a Mediterranean FAP and of other future programmes;

*raise* the awareness of and *inform* the public, particularly the young generation, on forest issues so they will be better appreciated by all;

*devise* ways of following up its recommendations, and *invite* FAO to inform intergovernmental bodies and the eleventh World Forest Congress thereon.

**Appendix 2**
**Agreement Adopted at the United Nations**
**Conference on Environment and Development**
**Rio de Janeiro, June 1992**

Non-legally binding authoritative statement of principles
for a global consensus on the management, conservation
and sustainable development of all types of forests

Preamble

(a) The subject of forests is related to the entire range of environmental and development issues and opportunities, including the right to socio-economic development on a sustainable basis.

(b) The guiding objective of these principles is to contribute to the management, conservation and sustainable development of forests and to provide for their multiple and complementary functions and uses.

(c) Forestry issues and opportunities should be examined in a holistic and balanced manner within the overall context of environment and development, taking into consideration the multiple functions and uses of forests, including traditional uses, and the likely economic and social stress when these uses are constrained or restricted, as well as the potential for development that sustainable forest management can offer.

(d) These principles reflect a first global consensus on forests. In committing themselves to the prompt implementation of these principles, countries also decide to keep them under assessment for their adequacy with regard to further international cooperation on forest issues.

(e) These principles should apply to all types of forests, both natural and planted, in all geographic regions and climatic zones, including austral, boreal, subtemperate, temperate, subtropical and tropical.

(f) All types of forests embody complex and unique ecological processes which are the basis for their present and potential capacity to provide resources to satisfy human needs as well as environmental values, and as such their sound management and conservation is of concern to the Governments of the countries to which they belong and are of value to local communities and to the environment as a whole.

(g) Forests are essential to economic development and the maintenance of all forms of life.

(h) Recognizing that the responsibility for forest management, conservation and sustainable development is in many States allocated among federal/national, state/provincial and local levels of government, each State, in accordance with its constitution and/or national legislation, should pursue these principles at the appropriate level of government.

## Principles/Elements

1.(a) "States have, in accordance with the Charter of the United Nations and the principles of international law, the sovereign right to exploit their own resources pursuant to their own environmental policies and have the responsibility to ensure that activities within their jurisdiction or control do not cause damage to the environment of other States or of areas beyond the limits of national jurisdiction."

(b) The agreed full incremental cost of achieving benefits associated with forest conservation and sustainable development requires increased international cooperation and should be equitably shared by the international community.

2.(a) States have the sovereign and inalienable right to utilize, manage and develop their forests in accordance with their development needs and level of socio-economic development and on the basis of national policies consistent with sustainable development and legislation, including the conversion of such areas for other uses within the overall socio-economic development plan and based on rational land-use policies.

(b) Forest resources and forest lands should be sustainably managed to meet the social, economic, ecological, cultural and spiritual human needs of present and future generations. These needs are for forest products and services, such as wood and wood products, water, food, fodder, medicine, fuel, shelter, employment, recreation, habitats for wildlife, landscape diversity, carbon sinks and reservoirs, and for other forest products. Appropriate measures should be taken to protect forests against harmful effects of pollution, including air-borne pollution, fires, pests and diseases in order to maintain their full multiple value.

(c) The provision of timely, reliable and accurate information on forests and forest ecosystems is essential for public understanding and informed decision-making and should be ensured.

(d) Governments should promote and provide opportunities for the participation of interested parties, including local communities and indigenous people, industries, labour, non-governmental organizations and individuals, forest dwellers and women, in the development, implementation and planning of national forest policies.

3.(a) National policies and strategies should provide a framework for increased efforts, including the development and strengthening of institutions and programmes for the management, conservation and sustainable development of forests and forest lands.

(b) International institutional arrangements, building on those organizations and mechanisms already in existence, as appropriate, should facilitate international cooperation in the field of forests.

(c) All aspects of environmental protection and social and economic development as they relate to forests and forest lands should be integrated and comprehensive.

4. The vital role of all types of forests in maintaining the ecological processes and balance at the local, national, regional and global levels through, inter alia, their role in protecting fragile ecosystems, watersheds and freshwater resources and as rich storehouses of biodiversity and biological resources and sources of genetic material for biotechnology products, as well as photosynthesis, should be recognized.

5.(a) National forest policies should recognize and duly support the identity, culture and the rights of indigenous people, their communities and other communities and forest dwellers. Appropriate conditions should be promoted for these groups to enable them to have an economic stake in forest use, perform economic activities, and achieve and maintain cultural identity and social organization, as well as adequate levels of livelihood and well-being, through, inter alia, those land tenure arrangements which serve as incentives for the sustainable management of forests.

(b) The full participation of women in all aspects of the management, conservation and sustainable development of forests should be actively promoted.

6.(a) All types of forests play an important role in meeting energy requirements through the provision of a renewable source of bio-energy, particularly in developing countries, and the demands for fuelwood for household and industrial needs should be met through sustainable forest management, afforestation and reforestation. To this end, the potential contribution of plantations of both indigenous and introduced species for the provision of both fuel and industrial wood should be recognized.

(b) National policies and programmes should take into account the relationship, where it exists, between the conservation, management and sustainable development of forests and all aspects related to the production, consumption, recycling and/or final disposal of forest products.

(c) Decisions taken on the management, conservation and sustainable development of forest resources should benefit, to the extent practicable, from a comprehensive assessment of economic and non-economic values of forest goods and services and of the environmental costs and benefits. The development and improvement of methodologies for such evaluations should be promoted.

(d) The role of planted forests and permanent agricultural crops as sustainable and environmentally sound sources of renewable energy and industrial raw material should be recognized, enhanced and promoted. Their contribution to the maintenance of ecological processes, to offsetting pressure on primary/old-growth forest and to providing regional employment and development with the adequate involvement of local inhabitants should be recognized and enhanced.

(e) Natural forests also constitute a source of goods and services, and their conservation, sustainable management and use should be promoted.

7.(a) Efforts should be made to promote a supportive international economic climate conducive to sustained and environmentally sound development of forests in all countries, which include, inter alia, the promotion of sustainable patterns of production and consumption, the eradication of poverty and the promotion of food security.

(b) Specific financial resources should be provided to developing countries with significant forest areas which establish programmes for the conservation of forests including protected natural forest areas. These resources should be directed notably to economic sectors which would stimulate economic and social substitution activities.

8.(a) Efforts should be undertaken towards the greening of the world. All countries, notably developed countries, should take positive and transparent action towards reforestation, afforestation and forest conservation, as appropriate.

(b) Efforts to maintain and increase forest cover and forest productivity should be undertaken in ecologically, economically and socially sound ways through the rehabilitation, reforestation and re-establishment of trees and forests on unproductive, degraded and deforested lands, as well as through the management of existing forest resources.

(c) The implementation of national policies and programmes aimed at forest management, conservation and sustainable development, particularly in developing countries, should be supported by international financial and technical cooperation, including through the private sector, where appropriate.

(d) Sustainable forest management and use should be carried out in accordance with national development policies and priorities and on the basis of environmentally sound national guidelines. In the formulation of such guidelines, account should be taken, as appropriate and if applicable, of relevant internationally agreed methodologies and criteria.

(e) Forest management should be integrated with management of adjacent areas so as to maintain ecological balance and sustainable productivity.

(f) National policies and/or legislation aimed at management, conservation and sustainable development of forests should include the protection of ecologically viable representative or unique examples of forests, including primary/old-growth forests, cultural, spiritual, historical, religious and other unique and valued forests of national importance.

(g) Access to biological resources, including genetic material, shall be with due regard to the sovereign rights of the countries where the forests are located and to the sharing on mutually agreed terms of technology and profits from biotechnology products that are derived from these resources.

(h) National policies should ensure that environmental impact assessments should be carried out where actions are likely to have significant adverse impacts on important forest resources, and where such actions are subject to a decision of a competent national authority.

9.(a) The efforts of developing countries to strengthen the management, conservation and sustainable development of their forest resources should be supported by the international community, taking into account the importance of redressing external indebtedness, particularly where aggravated by the net transfer of resources to developed countries, as well as the problem of achieving at least the replacement value of forests through improved market access for forest products, especially processed products. In this respect, special attention should also be given to the countries undergoing the process of transition to market economies.

(b) The problems that hinder efforts to attain the conservation and sustainable use of forest resources and that stem from the lack of alternative options available to local communities, in particular the urban poor and poor rural populations who are economically and socially dependent on forests and forest resources, should be addressed by Governments and the international community.

(c) National policy formulation with respect to all types of forests should take account of the pressures and demands imposed on forest ecosystems and resources from influencing factors outside the forest sector, and intersectoral means of dealing with these pressures and demands should be sought.

10. New and additional financial resources should be provided to developing countries to enable them to sustainably manage, conserve and develop their forest resources, including through afforestation, reforestation and combating deforestation and forest and land degradation.

11. In order to enable, in particular, developing countries to enhance their endogenous capacity and to better manage, conserve and develop their forest resources, the access to and transfer of environmentally sound technologies and corresponding know-how on favourable terms, including on concessional and preferential terms, as mutually agreed, in accordance with the relevant provisions of Agenda 21, should be promoted, facilitated and financed, as appropriate.

12.(a) Scientific research, forest inventories and assessments carried out by national institutions which take into account, where relevant, biological, physical, social and economic variables, as well as technological development and its application in the field of sustainable forest management, conservation and development, should be strengthened through effective modalities, including international cooperation. In this context, attention should also be given to research and development of sustainably harvested non-wood products.

(b) National and, where appropriate, regional and international institutional capabilities in education, training, science, technology, economics, anthropology and social aspects of forests and forest management are essential to the conservation and sustainable development of forests and should be strengthened.

(c) International exchange of information on the results of forest and forest management research and development should be enhanced and broadened, as appropriate, making full use of education and training institutions, including those in the private sector.

(d) Appropriate indigenous capacity and local knowledge regarding the conservation and sustainable development of forests should, through institutional and financial support, and in collaboration with the people in local communities concerned, be recognized, respected, recorded, developed and, as appropriate, introduced in the implementation of programmes. Benefits arising from the utilization of indigenous knowledge should therefore be equitably shared with such people.

13.(a) Trade in forest products should be based on non-discriminatory and multilaterally agreed rules and procedures consistent with international trade law and practices. In this context, open and free international trade in forest products should be facilitated.

(b) Reduction or removal of tariff barriers and impediments to the provision of better market access and better prices for higher value-added forest products and their local processing should be encouraged to enable producer countries to better conserve and manage their renewable forest resources.

(c) Incorporation of environmental costs and benefits into market forces and mechanisms, in order to achieve forest conservation and sustainable development, should be encouraged both domestically and internationally.

(d) Forest conservation and sustainable development policies should be integrated with economic, trade and other relevant policies.

(e) Fiscal, trade, industrial, transportation and other policies and practices that may lead to forest degradation should be avoided. Adequate policies, aimed at management, conservation and sustainable development of forests, including where appropriate, incentives, should be encouraged.

14. Unilateral measures, incompatible with international obligations or agreements, to restrict and/or ban international trade in timber or other forest products should be removed or avoided, in order to attain long-term sustainable forest management.

15. Pollutants, particularly air-borne pollutants, including those responsible for acidic deposition, that are harmful to the health of forest ecosystems at the local, national, regional and global levels should be controlled.

[A/CONF.151/6/REV.1: FOREST PRINCIPLES. Distr.: General. Original: English]

# Contributors

**Herman E. Daly**
World Bank
Environment Department
Washington, D.C.

**Richard A. Houghton**
Woods Hole Research Center
Woods Hole, Massachusetts

**Jagmohan S. Maini**
Forestry Canada
Hull, Quebec
Canada

**Darrell A. Posey**
Amazon Institute of Ethnobiology
Belém
Brazil

**Kilaparti Ramakrishna**
Woods Hole Research Center
Woods Hole, Massachusetts

**Robert Repetto**
World Resources Institute
Washington, D.C.

**Ola Ullsten**
Embassy of Sweden
Rome
Italy

**George M. Woodwell**
Woods Hole Research Center
Woods Hole, Massachusetts

# MEDWAY CAMPUS LIBRARY

This book is due for return or renewal on the last date stamped below,
but may be recalled earlier if needed by other readers.
Fines will be charged as soon as it becomes overdue.

Y

NSTITUTE
Central Avenue,
Chatham Maritime,
Chatham,
Kent ME4 4TB.

*the*
**UNIVERSITY**
*of*
**GREENWICH**